Global Conference on Process Safety and Big Data 2022

AA009141

Frankfurt am Main, Germany
11 – 12 October 2022

ISBN: 978-1-7138-6321-2

Printed from e-media with permission by:

Curran Associates, Inc.
57 Morehouse Lane
Red Hook, NY 12571

Some format issues inherent in the e-media version may also appear in this print version.

Printed with permission by Curran Associates, Inc. (2023)

For permission requests, please contact AIChE
at the address below.

AIChE
120 Wall Street, FL 23
New York, NY 10005-4020

Phone: (800) 242-4363
Fax: (203) 775-5177

www.aiche.org

Additional copies of this publication are available from:

Curran Associates, Inc.
57 Morehouse Lane
Red Hook, NY 12571 USA
Phone: 845-758-0400
Fax: 845-758-2633
Email: curran@proceedings.com
Web: www.proceedings.com

TABLE OF CONTENTS

KEYNOTE SESSION 1

PANEL 1: CHALLENGES TO INTEGRATE NEW DIGITALIZATION TOOLS & TECHNOLOGIES INTO EXISTING FACILITIES

SESSION 1: AUGMENTED REALITY & ROBOTIC

SESSION 2: BIG DATA ANALYTICS FOR PROCESS SAFETY

KEYNOTE SESSION 2

PANEL 2: THE JOURNEY FOR SUSTAINABLE AND INNOVATIVE IMPLEMENTATION OF PROCESS SAFETY BIG DATA ANALYTICS PROJECTS

SESSION 3: PREDICTION OF MAINTENANCE AND OPERATIONAL ERRORS

SESSION 4: TOOLS TO AVOID HUMAN ERRORS

POSTER SESSION 1

POSTER SESSION 2

the human Chevron energy company™

Data for Safeguard Effectiveness

Safeguards

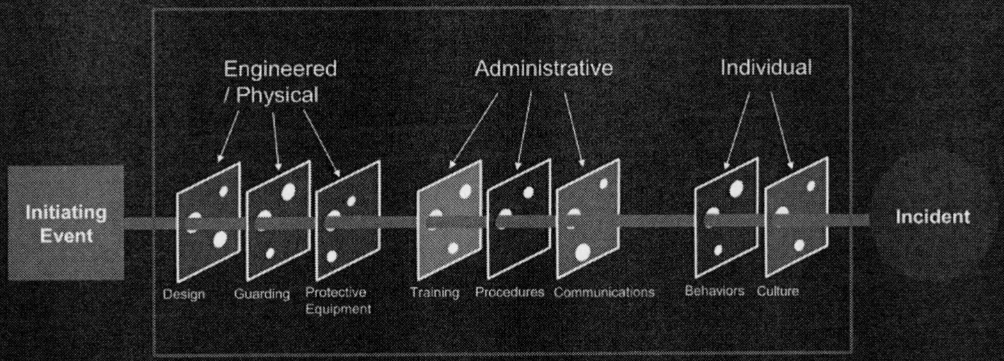

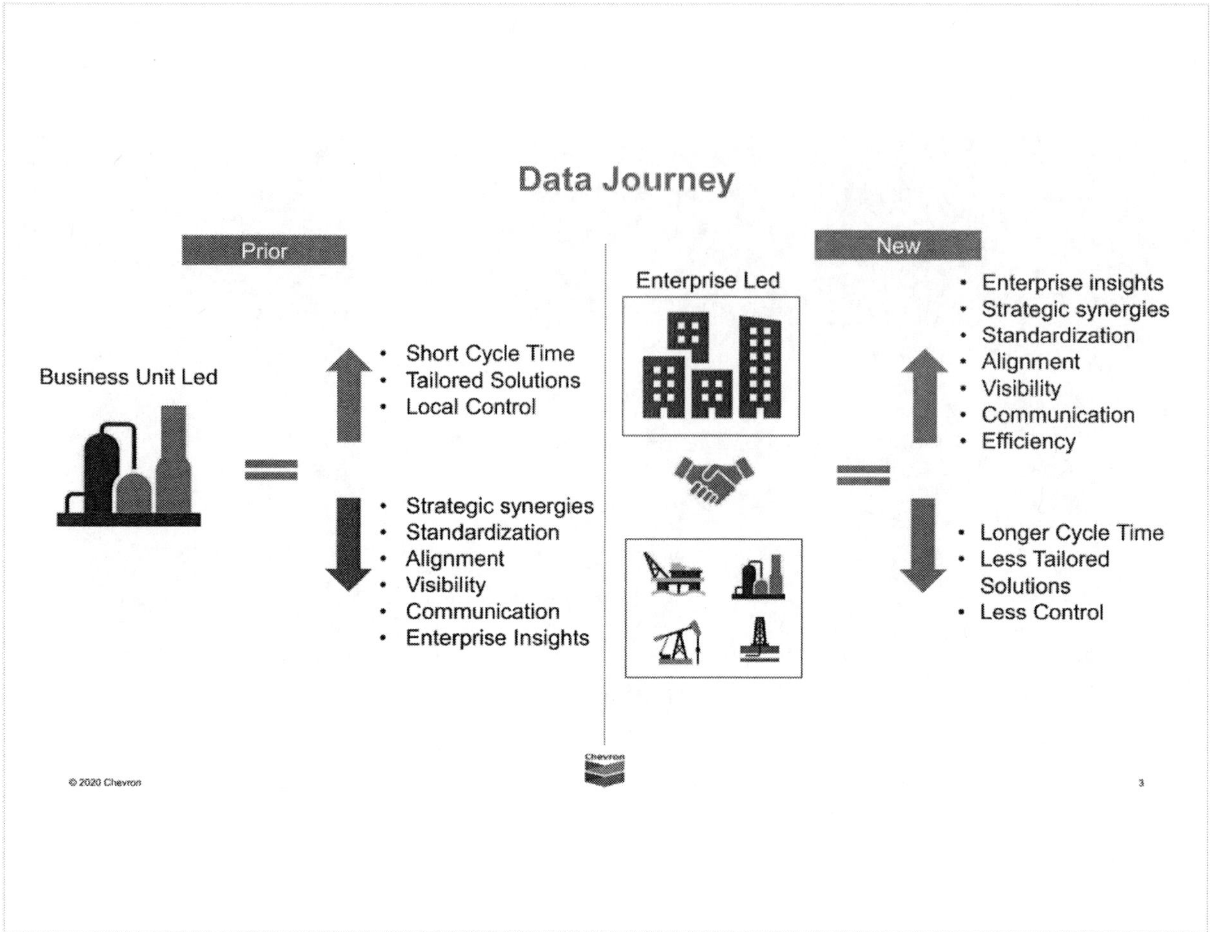
Data Journey
Prior
New
Business Unit Led
• Short Cycle Time
• Tailored Solutions
• Local Control
• Strategic synergies
• Standardization
• Alignment
• Visibility
• Communication
• Enterprise Insights
Enterprise Led
• Enterprise insights
• Strategic synergies
• Standardization
• Alignment
• Visibility
• Communication
• Efficiency
• Longer Cycle Time
• Less Tailored Solutions
• Less Control
© 2020 Chevron
Chevron
3

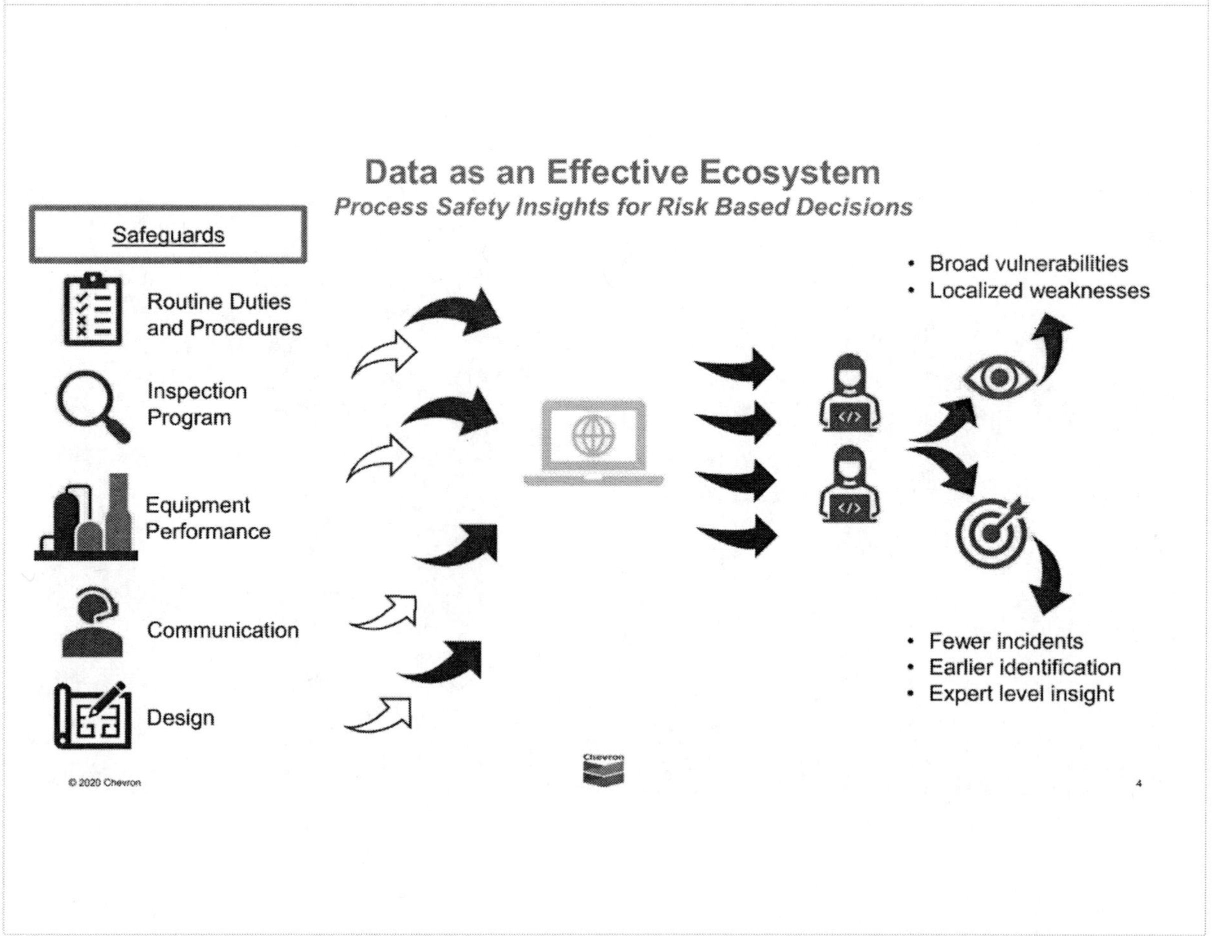
Data as an Effective Ecosystem
Process Safety Insights for Risk Based Decisions
Safeguards
Routine Duties and Procedures
Inspection Program
Equipment Performance
Communication
Design
• Broad vulnerabilities
• Localized weaknesses
• Fewer incidents
• Earlier identification
• Expert level insight
© 2020 Chevron
Chevron
4

Real World Impact

Safeguards	Information type	Situation	Impact
Routine Duties and Procedures			
Inspection Program	• Inspection results • Equipment life calculations • Trend Analysis	Determine equipment life	• Manual Inputs • Manual data quality check • Manual calculate vessel integrity • Review for issues • Take action •Update inspection frequency •Remove from service
Equipment Performance	• Operating data • Trend analysis • Failure analysis	Across our global portfolio, learn more quickly and effectively on how to improve equipment reliability and integrity	• Partial data infrastructure • Partial SME support • Limited systemic identification and tracking
Communication			
Design			

How do you this?

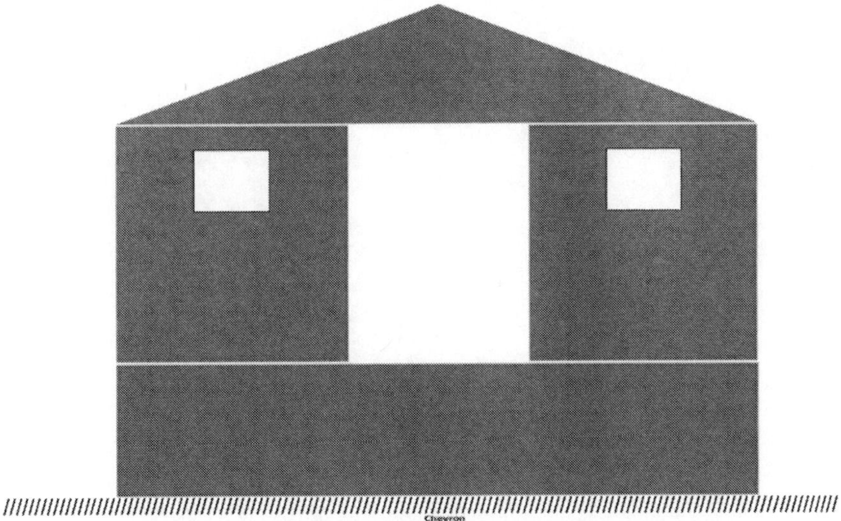

The Foundation

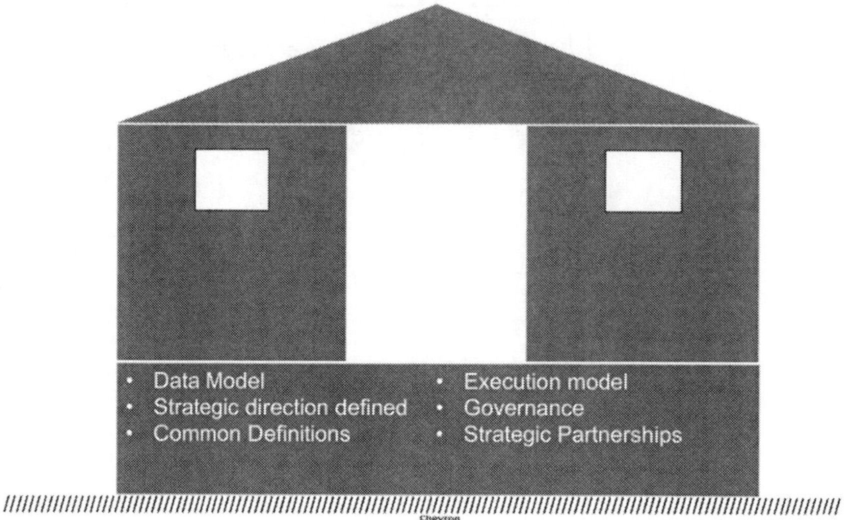

© 2020 Chevron

7

The Walls

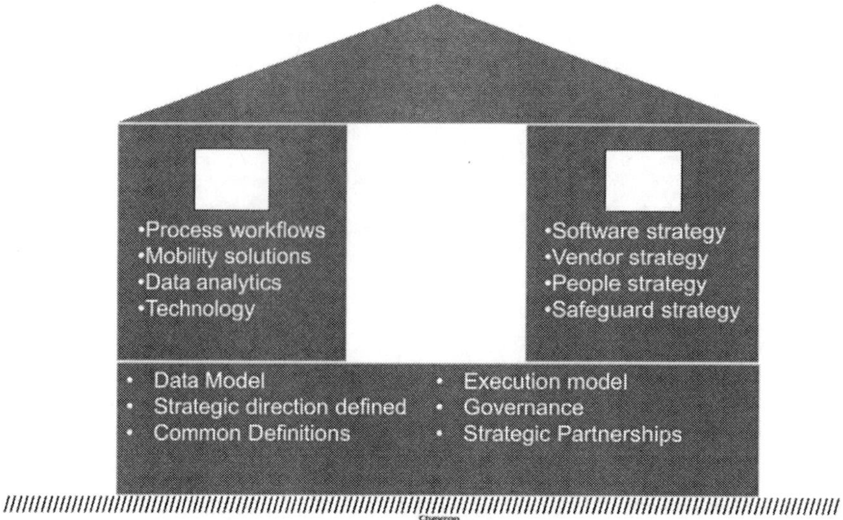

© 2020 Chevron

8

The Roof

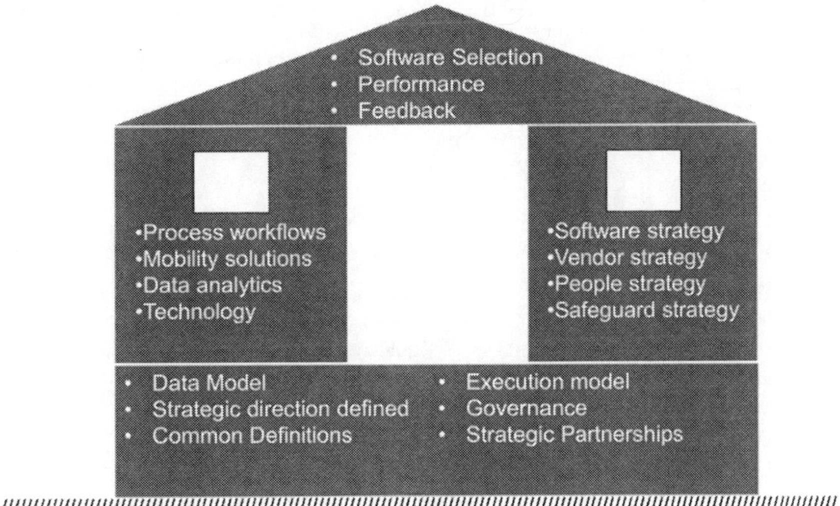

The Re-bar

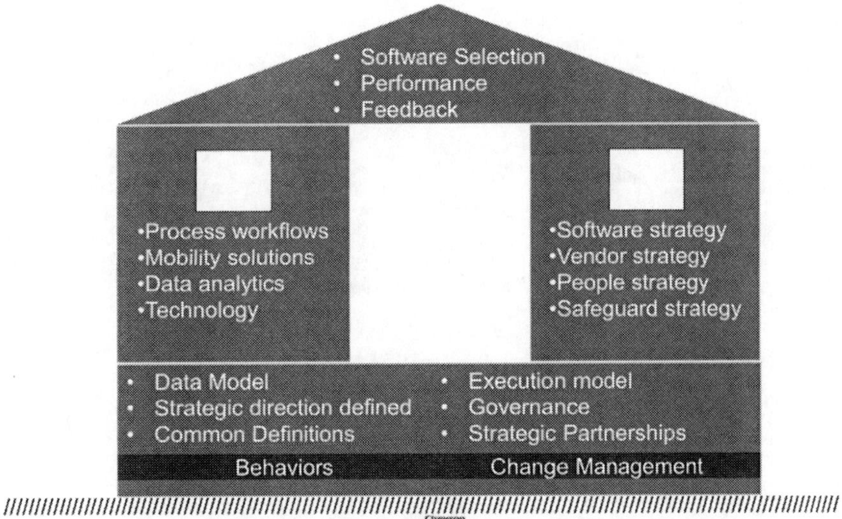

Bringing it all Home

Safeguards
- Routine Duties and Procedures
- Inspection Program
- Equipment Performance
- Communication
- Design

Information type
- Inspection results
- Equipment life calculations
- Trend Analysis

- Operating data
- Trend analysis
- Failure analysis

Situation

Determine equipment life

Across our global portfolio, learn more quickly and effectively on how to improve equipment reliability and integrity

Impact
- Input, QC, calculations are automated
- System flags concerns
- People only review what is necessary
  - Automate frequency changes and maintenance work request

- Full data infrastructure
- Live Data
- Remote SME analysis
- Global trending
- Non-conformance tracking and action

Summary

Safeguards
- Routine Duties and Procedures
- Inspection Program
- Equipment Performance
- Communication
- Design

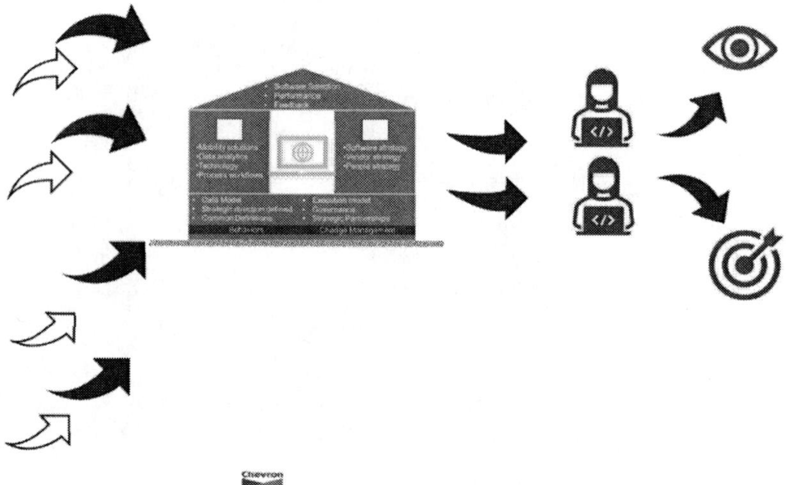

the
human Chevron energy
company™

End

2021 EUROPEAN CONFERENCE ON PROCESS SAFETY & BIG DATA

CCPS EPSC

Predictive Tools

Spill Prevention

PPG

PPG EHS Management System

- 12 Elements
- 61 requirements
- Plus supporting documentation

EHS Management System:

- Fulfilling requirements for ISO14001, ISO18001 and ISO 45001 certification
- Driving incident reduction
- Addressing regulatory compliance
- Providing platform for EHS operational excellence

Assessing EHS maturity:

- Self-assessments of elements and requirements at local level
- Calibration at country/regional level
- Formalised auditing process at global level

PPG

Incident Risk Categorisation

Process:

- Scoring system for elements and requirements, focused on documentation, interviews and observations
- Use of **Bradley Curve™** for categorization
- Incident risk categorisation of locations based upon scoring of specific elements and requirements
  - Management leadership
  - Equipment Integrity
  - Ergonomics
  - Powered Industrial Trucks
  - Risk Assessment & management
  - Slip, Trip & Fall Prevention
  - Spill Elimination

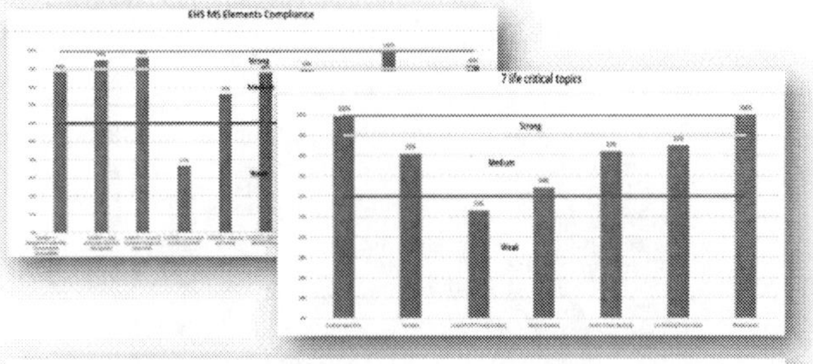

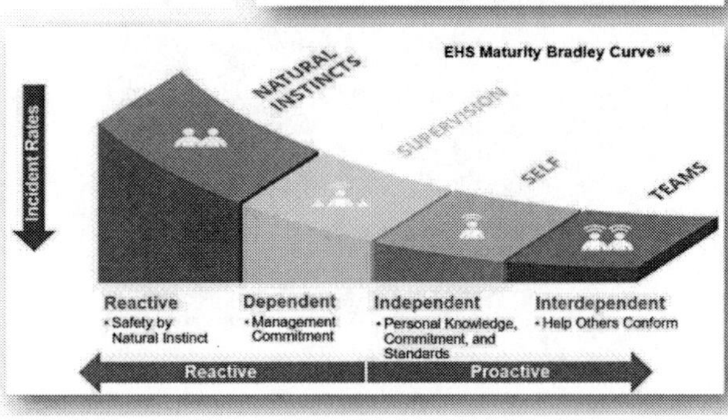

Context and Scope of Work

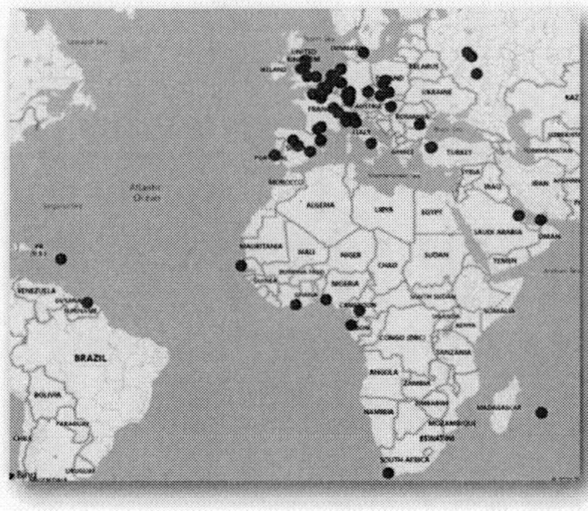

EMEA Region

- More than 70 sites
- Spill Elimination applies to approximately 60 sites
- Over 29 countries
- Speaking 18 different languages
- Recent acquisitions out of scope

Highly Complex Environment

PPG EHS Management System

KEY REQUIREMENTS:

1. All applicable facilities shall establish a Spill Elimination Program, including the program steps of
   a. Assessment of facility spill elimination performance.
   b. Improvement plans based on prioritized assessment of risk.
   c. Corrective action plan with defined dates.
   d. Completion of planned actions.
2. Spill performance and the Spill Elimination Program are the responsibility of the facility Leadership Team.

Plants required to complete Spill Assessment once a year

Monitoring performance without adding complexity

- Simple, quick and effective survey process
  - All locations need to complete the survery
  - Survey received by email (SurveyMonkey form)
  - 10 questions to answer, with an estimated time for completion of 3 minutes
  - Survey completed by EHS manager based in the location
- Questions selected to determine spill likelihood based on historical incidents and manufacturing technology
- Answers determine the action plan
- All inputs are consolidated at regional level and presented on a quarterly basis to location and business leaders
- The survey does not replace the assessment of the Spill Elimination requirement of the PPG EHS Management System

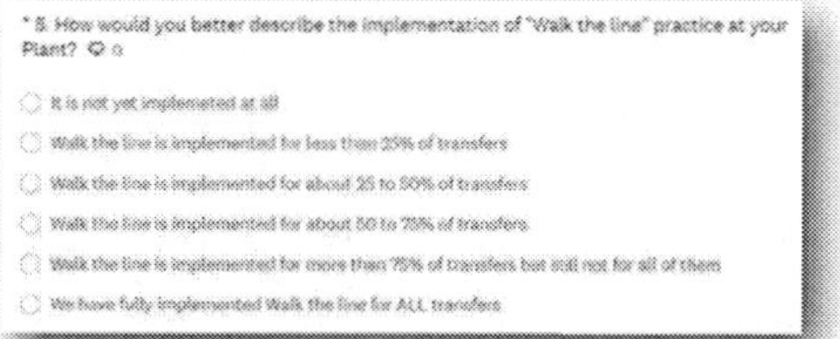

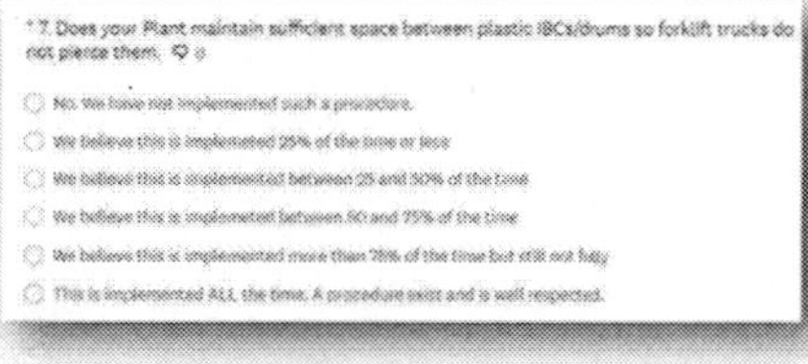

Key drivers for spill prevention at PPG

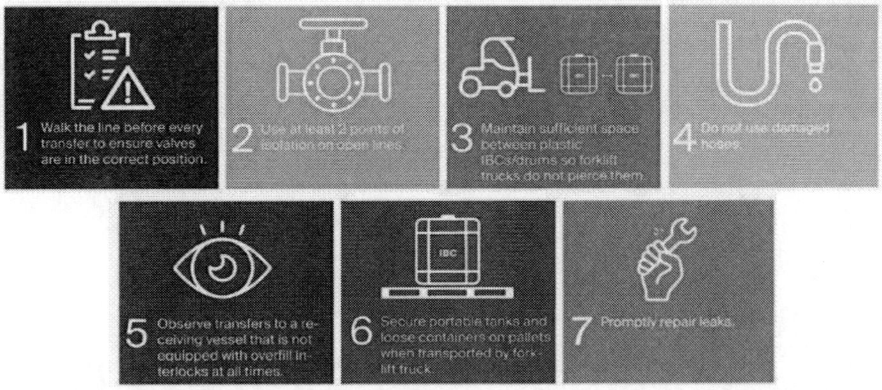

Drivers selected based upon historical incidents in PPG

Other companies might select different drivers

Scoring and ranking process

- The results of the survey were compared with the last 3 years performance
- Drivers are weighted differently and may change based upon types of incident
  - Walk the line & Observe transfers do have higher weight / correlation
  - Rest of the drivers do have lower weight / correlation
- The **Spill Likelihood Index** reflects the maturity of the locations and ranks locations in 4 different levels of risk (lower score, more risk)

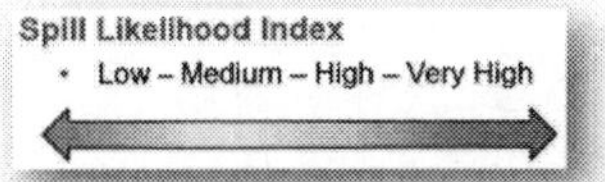

- Locations in 'Very High' and 'High' level do require special focused improvements and training
- Index can be calculated by Location, Country, Region and Business and monitored to achieve continuous improvement

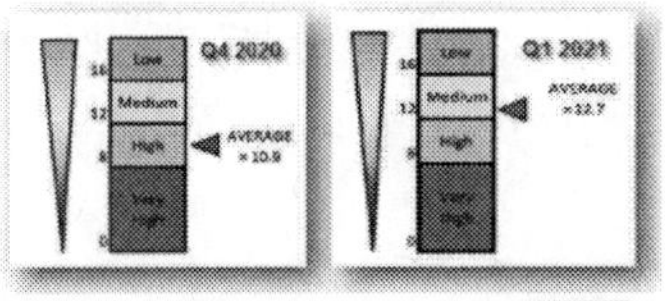

Alignment / Correlation with safety performance

Spill Likelihood Index

- Low – Medium – High – Very High

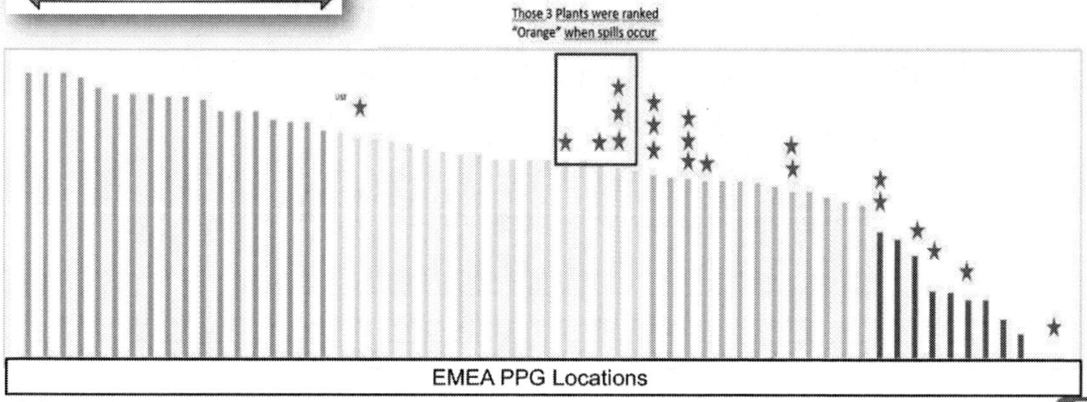

★ Reportable spill during last 3 years

PPG

Spill Assessment Tool vs Spill Likelihood Index

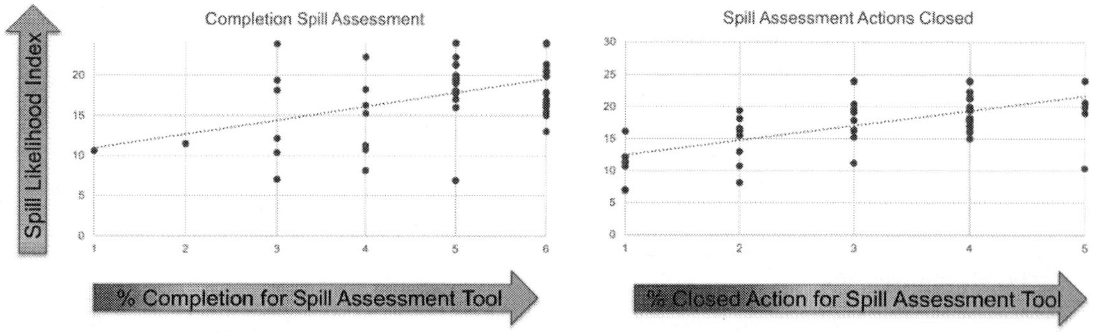

Completion of Spill Assessment Tool is aligned with High Spill Indexes

PPG

Overall Region performance can be monitored Quarterly

	2020 Q4			
Nº Plants	17	17	15	12
% of Plants	27%	28%	25%	20%

0 8 12 16 — 10.9

	2021 Q1			
Nº Plants	13	12	17	17
% of Plants	22%	20%	29%	29%

0 8 12 16 — 12.7

	2021 Q2			
Nº Plants	10	13	18	18
% of Plants	17%	22%	31%	31%

0 8 12 16 — 13.4

	2021 Q3			
Nº Plants	3	7	12	38
% of Plants	5%	12%	20%	63%

0 8 12 16 — 17.7

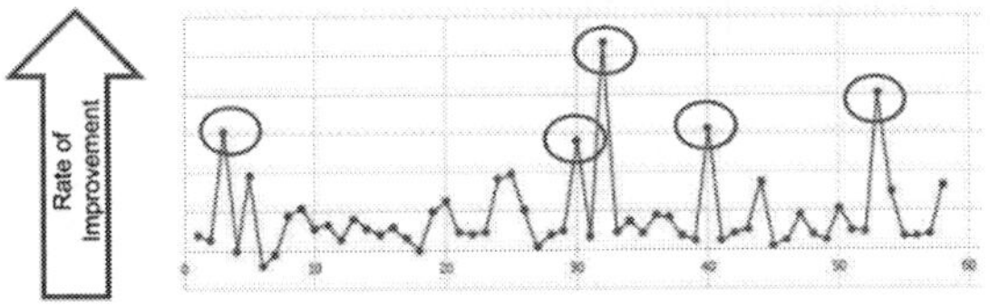

Outliners can easily be detected and asked to present evidence of improvement

PPG

Can be used as Continuous Improvement Tool

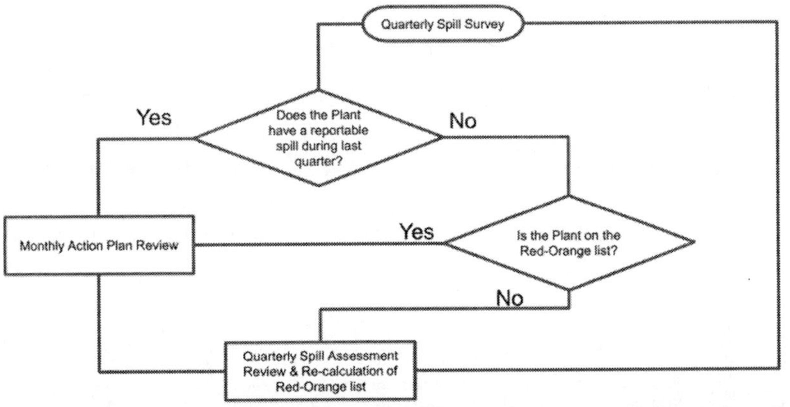

Action plans for Red & Orange locations are monitored till they reach Yellow or Green Rank.

PPG

Next Steps

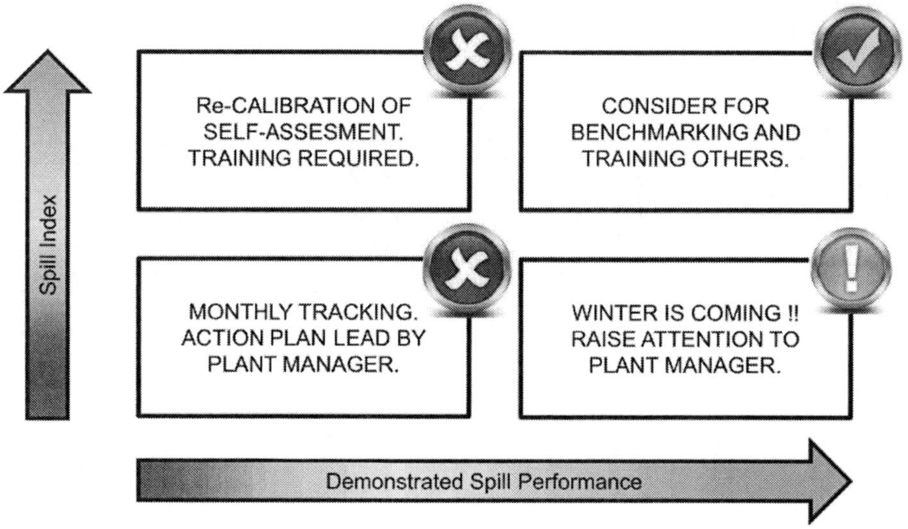

PPG

Learning and conclusions

- Important to understand well the root cause, frequency and historical trend for incidents
- Simple collection of data is more than enough to drive continuous improvement
- Ranking locations on maturity levels helps to focus resources on training and improvements
- Monitoring progress promotes best practice sharing… and healthy competition

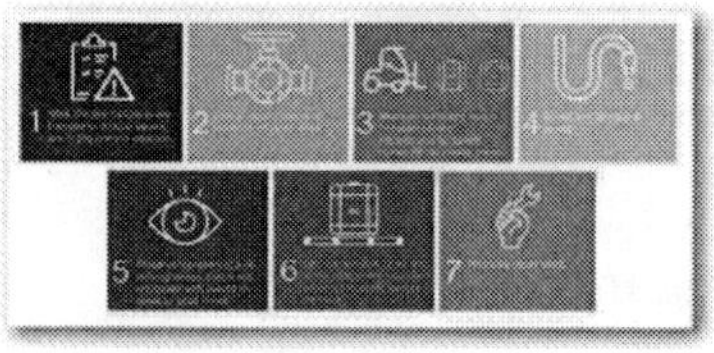

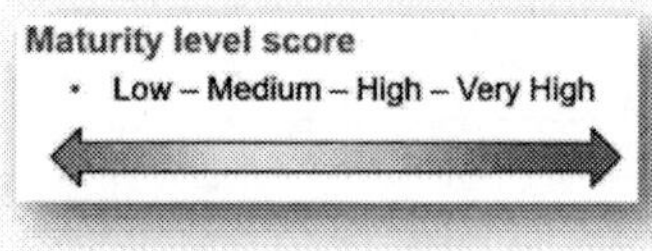

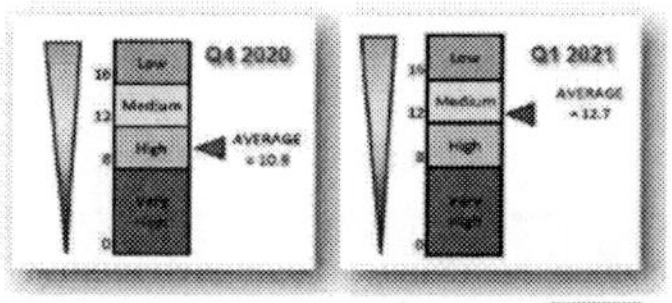

PPG

2021 EUROPEAN CONFERENCE ON PROCESS SAFETY & BIG DATA
CCPS EPSC

3:20 – 4:10 PM Panel Session Chairs & Moderator: *Joe Stough, Arthur Groot & Tijs Koerts*	**PANEL 1:** Challenges to Integrate New Digitalization Tools & Technologies into Existing Facilities **Panelist:** • ***Wilfred Beijers***, Chevron • ***Greg Nesmith***, Dow • ***Michel Lutz,*** TotalEnergies • ***Awad Shehri***, Saudi Aramco

October 13, 2021

Slide 1

the human Chevron energy company™

HSE Digital Platform

Wilfred Beijers
October 2021

Chevron HSE IT – Our Journey

- During the late 1990's Chevron developed many **home grown IT applications** to support their disparate HSE processes
- In the early 2000's Chevron started to **standardize HSE processes**, and as a result were able to develop more **global standardized IT applications** – the motto became *Buy over Build*
- Currently, we start to take advantage of **low-code applications**
  - Low code applications are being developed to interact with our **Data Lake**
  - Data can be collected or manipulated through low-code apps
  - Low code reporting tools allow central teams to focus on the data, while end users or their business units can maintain reporting tools locally
- Starting in 2020, Chevron reorganized their IT organizations and set up 13 **Digital Platforms**
  - Chevron now uses **Agile** as their software development methodology
  - The Digital Platforms are also staffed in an agile manner through **Chapters**
  - Chevron expanded their partnership with **Managed Service Providers** to support their applications

Chevron Digital Platform Architecture

Our guardrails ensure we design platforms in a standard way, enabling reuse, scalability and integration

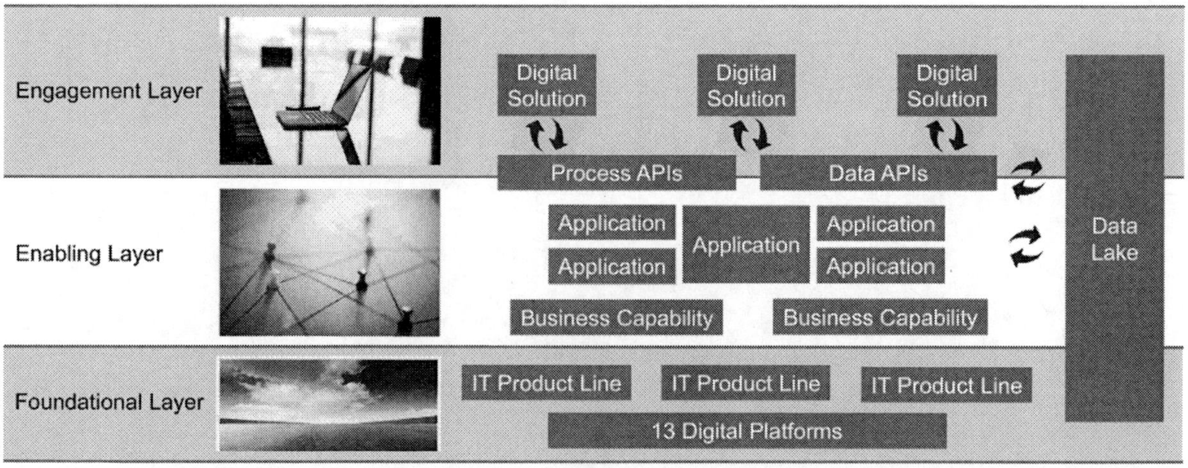

Summary

- Chevron is converting its old architecture into a **Technology Ecosystem** that is managed by 13 Digital Platforms
- **Platforms autonomously manage their applications and data**, but collaborate with other Platforms
  - Platforms consist of Product Lines that support **Products and Business Capabilities**
- Chevron's **Data Lake is becoming the central hub** for sourcing and provisioning data
  - In addition to reporting data, the Data Lake includes **Master Data, Snapshots**
- Chevron is taking advantage of **low-code technical solutions**
- Applications are expected to interact with the Data Lake
  - APIs support a multitude of **interfaces**
  - **Reporting, dashboards and analytics** are sourcing directly from the Data Lake
- Additional integrations being set up to collect data through IIoT devices

5

PORTABLE DEVICES AND HAZARDOUS AREA CLASSIFICATION

WHAT TO CONSIDER AS WE ACCELERATE THEIR USE

Greg Nesmith, PE, CFSE

Senior Process Safety Manager for Industry 4.0 at Dow

Responsibilities

- Lead structured risk analyses
- Implement innovative solutions for Process Safety challenges
- Identify failure modes in new technologies
- Implement global process safety training

Link between Process Safety and Big Data

- Apply key instrument and automation standards (e.g., IEC 61511) in Dow
- Lead Data Analytics for SIS project

Career Highlights

- 31 years with Dow Chemical
- In Process Safety since 2007
- Certified Functional Safety Expert in 2017

email: gregnesmith@dow.com

THE PROBLEM

There have been an increasing number of near misses recently in which portable electronic devices were either approved or used in classified process areas without proper certification, training, and/or operating discipline.

We must act to eliminate these issues before there is an incident.

The Solution

- **Awareness: Learning experiences are being more widely communicated the challenges to everyone.**
- Resources: The number of resources qualified to assess devices and create the appropriate training is limited. Active use of these devices must be limited until classification and training are complete.
- **Teamwork: Since there are many different groups approving this technology, we must coordinate their efforts.**

- Process Safety: Engage in your business and geography.

DOW

9

Hazardous Area Classification

Organizations and stakeholders increasing the use of mobile devices and technologies in process areas:

- Operations – Laptops, Tablets, Phones, T/A tracking systems
- Construction and Maintenance – Phones and Radios, test equipment and tools
- Robotics – UAV's (Drones), Robots, Cameras
- Industrial Hygiene – Hearing Aids, Noise Cancelling Headphones, etc.
- Health & Safety – Exoskeleton, Covid Proximity Monitoring Devices, etc.
- Experimental Technologies – RealWear, Cameras, Monitoring Devices
- Others?

Significant needs:

1. **Are qualified experts engaged?**
2. **Is proper operating discipline and training utilized?**
3. **Are we collaborating to optimize preferred platforms and tools?**

DOW

10

Hazardous Area Classification

1. Are qualified experts engaged?

Question: What qualifies a person to select and specify electronic equipment for use in a hazardous (classified) location?

Answer:

- Awareness and knowledge of HAC processes is not enough.
- Qualified inspectors are required. In some cases, the expert must be a registered Professional Engineer.
- Devices suitable for hazardous service require a Nationally Recognized Testing Laboratory (NRTL) or Notified Body (NOBO) to inspect and certify the device.

All regulatory bodies require personnel selecting and specifying equipment for service in hazardous (classified) locations to be qualified.

DOW

Hazardous Area Classification

2. Is proper operating discipline and training utilized?

For Listed Portable Devices, two key provisions of the certification bind the user to define operating discipline and/or training to meet the certifier's conditions for supporting the HAC markings:

1) For all devices or equipment, the listing
the guidelines of the manufacturer's ins

WARNING

To reduce the risk of igniting an explosion which, if not avoided, could result in serious injury or death:
• Ensure that the 3M™ PELTOR™ WS™ LiteCom Pro III Headset and any intrinsically safe accessories are only used and stored in the classified areas consistent with the marked equipment ratings.
• Only use 3M™ PELTOR™ Authorized Service Centers for service and repair.

- **WARNING:** To prevent ignition of a hazardous atmosphere, batteries must only be changed or charged in an area known to be non-hazardous.

connector) in Hazardous area.

DOW

Hazardous Area Classification

2. Is proper operating discipline and training utilized?

2) For all portable devices, every certifier includes acceptable use conditions that define the conditions that meet the allowances for the HAC markings. Below are some examples.

Dell 7220EX | RealWear | Sonim XP8 | Ecom Tab-EX

The user instructions and the certifier's restrictions for use must be communicated and understood by the user.

DOW 13

Hazardous Area Classification

3. Are we collaborating to optimize preferred platforms and tools?

- Many devices, such as the RealWear portable computing system are best optimized when coupled with cell phone or tablet.
- Construction and Maintenance need cell phones or radios for communication.
- Industrial Hygiene wants noise cancelling headphones and ear plugs, which we want to couple by bluetooth to cell phones or tablets for communication in noisy areas.

Where we need common devices, reducing variability:

- Standardizes operating discipline and training
- Reduces administration, management, and costs
- Allows better purchasing opportunities for preferred devices
- Instills common knowledge among users

DOW 14

SUMMARY

What can we do?

- **Engage your Process Safety personnel.** Remind them of the significance of using suitable devices in hazardous locations. It is likely they are aware of the new technologies in their areas, but they may not be aware it is a Process Safety issue.
- **Find out who is selecting electronic devices**. Ask them who is providing qualified electrical and electronic device technical oversight.
- **Create some level of oversight.** If you do not have the expertise in your organization, make someone the conduit for oversight.
- **Discourage impulse purchases of mobile devices** without proper organizational controls.
- **Encourage Management of Change (MOC)** for even small changes and the inclusion of a Process Safety reviewer. Changing device models, going into different process areas, adding new users could all be examples of changes.

DOW

15

2021 EUROPEAN CONFERENCE ON PROCESS SAFETY & BIG DATA

CCPS EPSC

"Data transformation and Digital Factory at TotalEnergies"

Michel Lutz, Group Data Officer & Digital Factory Head of Data - @LutzAnalytics

October 13, 2021

Slide 16

aramco

Awad Al-Shehri

Is a Loss Prevention Unit Head working with Saudi Aramco overseeing safety support provided for hydrocarbon facilities.

He has more than 15 years of experience in the oil & gas industry and a Certified Safety Professional (CSP).

Currently he is a member of a corporate team leading digital transformation of process safety and operations management business processes.

aramco

Challenges to Integrate New Digitalization Tools & Technologies into Existing Facilities

2021 Virtual European Conference on Process Safety and Big Data (10/13-14/2021)
Center for Chemical Process Safety (CCPS) / European Process Safety Centre (EPSC)

Awad Al-Shehri, CSP
Saudi Aramco

Digitalization in Process Safety – Framework

aramco

- WHY DIGITALIZE? IS IT NECESSARY? WHAT CAN GO WRONG?
- Structured Approach for Successful Adoption
  - People, Process & Technology (PPT)
    - Balanced framework
    - Independent Action but combined Effect
    - Culture dependent

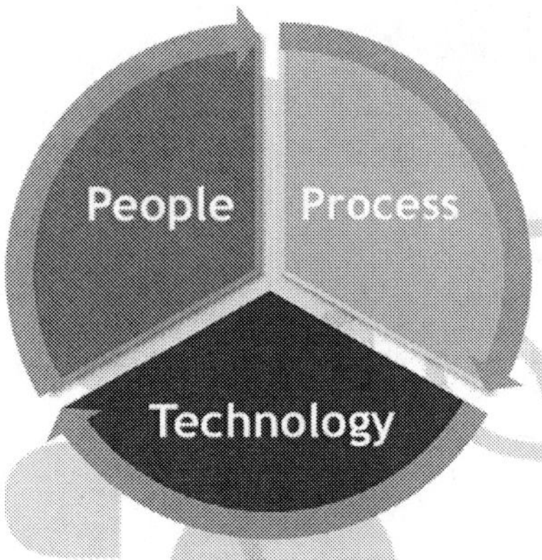

People

aramco

- **Workforce Skills**
  - Train users "new world requires new skills"
  - Watch for "technology over reliance" and unintended consequences
  - Encourage workforce skills diversification
- **Technology Fatigue**
  - Manage ongoing initiatives
  - Eliminate Overlapping
- **Project Marketing**
  - Promote project added value
  - Ensure stakeholders buy-in
  - Sustain management/users engagement
- **Communication**
  - Implement effective communication channels with end users
  - Encourage Two-way communication

Process

- **Technology Program**
  - Analyze potential high impact areas for improvement
  - Identify and select appropriate technology
  - Account for operations/customer impact
  - Oversee resources (beyond financial aspect)
- **Lessons Learned**
  - Cultivate learning from previous projects experiences
  - Capture successful resolutions of challenges
- **Measured Outcome**
  - Identify success measurements/KPIs
  - Frequently review/adjust
- **Management of Change**
  - Utilize effective MoC process throughout project lifecycle
  - Manage users expectations

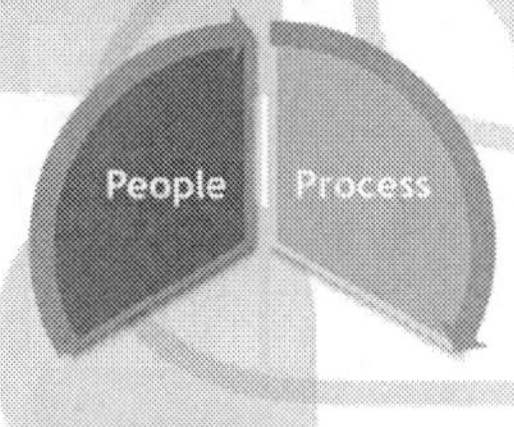

Technology

- **Integration with Legacy Systems**
  - Systems integration is key for interdependent tasks
  - Invest in your infrastructure and technology enablers
- **Rate of Obsoleteness**
  - Identify right technology with upgrade capability
  - Remember, new technology not always the right investment
  - Always keep a "look ahead"
- **Cybersecurity/System availability**
  - Assess IT risks and implement mitigations
  - Identify alternatives
- **Devices Conformance**
  - Ensure acceptable safety requirements
  - Assess devices for proper area classification

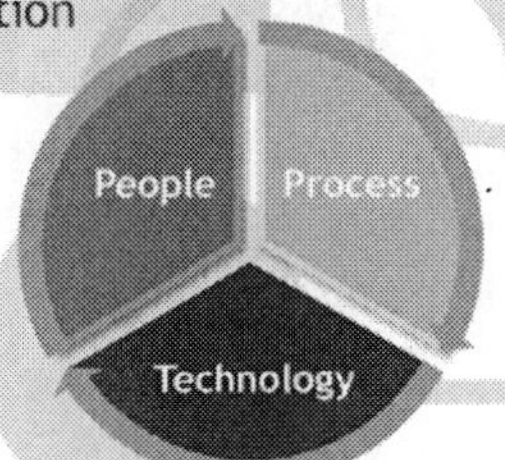

Success Story – Uthmaniyah Gas Plant

aramco

- One of the world's **Largest** gas processing plants
- One of the company **Oldest** gas plants
- Recognized by the 2019 World Economic Forum as
  - "Lighthouse" manufacturing facility
  - A leader in 4IR technology applications

New Initiative – Transforming Process Safety on a Grand Scale

aramco

- Digitalize several operational and safety business processes
  - Build "big data" collection database
- Integrate risk management and analysis tools
  - Enable AI for decision making
- Provide mobility solution
- Produce real time KPIs and Deviations Management platforms
- **Enhance Compliance, Reduce Human Errors and Improve Efficiency**

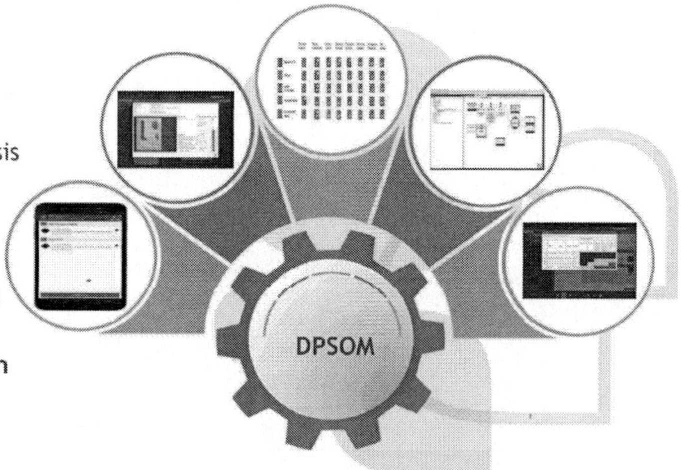

2021 EUROPEAN CONFERENCE ON PROCESS SAFETY & BIG DATA
CCPS EPSC

Panel 1
Q&A

October 13, 2021

Slide 26

VIRTUAL

2021 EUROPEAN CONFERENCE ON PROCESS SAFETY & BIG DATA

CCPS EPSC

Augmented Reality To Improve Process Safety

Hervé VAUDREY / Juhani LASSILA - DEKRA

DEKRA

DEKRA

GOALS FOR TODAY's TALK

- Share some thoughts and challenges about **improving Process Safety performance with Virtual Reality**
- Present a Virtual Reality solution from DEKRA to improve process safety:
  - A **Working** Application
  - **Cost-effective** & easy to implement
- Give some ideas of **possible applications**

DEKRA

DIGITALIZATION PS APPLICATIONS

Strong excitement about digitalization in multiple areas

- See the program of this conference
- Digital twins, 3D+ models, Virtual Reality, Internet of Things, ...

Challenges

- Some tools like Digital Twins are somewhat out-of-reach for many companies / SMEs
- Need significant amount of development / data
- Need specifically trained personnel

A control room in 2018

DEKRA

DEKRA VALTO 360°

- A Virtual Reality platform :
  - Based on 360° shootings in plants - even in HazLoc environments
  - Intuitive navigation & editing
  - Multi-platform : PC, tablet, mobile
  - Online 24/7 & Cyber-secure
  - Additions of **points of interest** of any type: images, documents, measurements, videos, diagrams, ...
- Recent additions :
  - Layers (eg HazLoc/ATEX classification) & connected devices/IOT
  - Task & file managers
  - Group based user & permission controls

DEKRA

CASE STUDY #1 : Steel Company (Finland)

Project description :

- The steel company is an historical client of DEKRA Industrial Inspection services
- They wanted to have better access to their inspection data & streamline preparation of inspections & improve training

Solution :

- The digital twin was built by shooting the facility over a couple of days
- Inspection points & historical data were loaded in the VR tool environment

Benefits :

- Faster location identification
- Less downtime
- Faster & more efficient training for "newcomers"

DEKRA

CASE STUDY #2 : Hospital Company (France)

Project description :

- The company owns 10+ hospitals in France and wanted to have a centralized view of the management of their fire equipment's

Solution :

- The digital twin was built by importing 3D model
- Fire safety equipment's (extinguishers, sprinklers, ...) locations & data were loaded in the VR tool environment for a pilot facility as "points of interest"

Benefits:

- Total oversight of fire safety 'systems' across the hospital – including automatic SMS/email alerts and reminders for inspections
- Pilot is being rolled out : "A beautiful tool perfectly targeted to give feedback / give a vision to the Management of the situation, to the Technical Director"

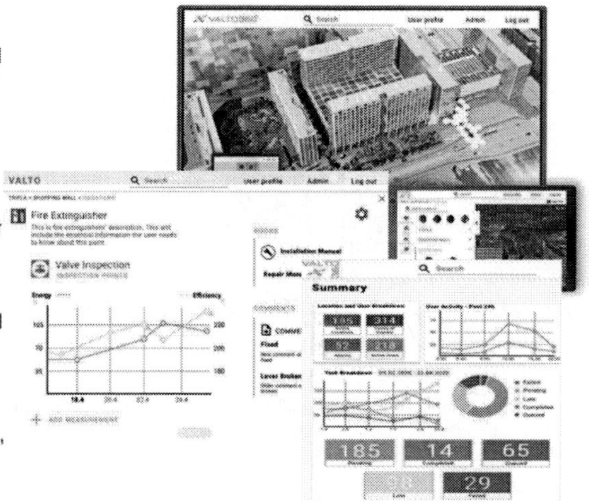

VR tool demo

DEKRA

Live demonstration in a petrochemical platform environment

DEKRA

PROCESS SAFETY APPLICATIONS OF VR TOOL

- **Visualization / Management of Process Safety Information**
  - For instance Pressure Safety Valves & Rupture Disks
- **Education / Trainings :**
  - To Procedures or Emergency Planning
  - To Safety Critical Elements
  - For Line breaking preparation
- **Support for Process Hazard Analysis**
- **Preparation of Process Safety Audits**
- **Visualisation of Hazardous Area Classification (Layers)**
- **Forensic studies / Incident investigation**
- …

Applicable across sectors and from labs to large scale facilities

Shoot once
Apply many times

A time-saver & safety performance enabler

DEKRA

About the speakers

Hervé Vaudrey

VP Global Sales Management

DEKRA Corporate

Member CCPS & EPSC

A passionate process safety professional with extensive chemical and pharmaceutical sector experience.

Experienced lecturer in a wide range of process safety subjects in the last 20 years. He has given over 100 process safety trainings worldwide (France, UK, Spain, Netherlands, India, China) in French, English and Spanish

He's currently VP Global Sales at DEKRA Corporate looking after key accounts of the process industries.

Contact information :
herve.vaudrey@dekra.com
www.dekra-process-safety.com
Phone : +33 6 76 47 15 65

Juhani Lassila

Valto360° Product Manager

DEKRA Finland

An enthusiastic digital product manager with background in design, marketing and analytics.

He's currently product manager of the digital platform VALTO360° and supporting other digital initiatives with Dekra SD.

Contact information :
juhani.lassila@dekra.com
www.dekra.com
Phone : +358 44 737 6977

How Covestro improves Safety and Reliability using Virtual Reality

Anke Köster – Global Qualification Management
Pietro Valsecchi – Digital R&D

Performance Materials for a Sustainable Growth

Serving key industries	Present around the globe	Technologically leading
Products and solutions for a diverse range of end-markets	Major sites in Europe, Asia and the Americas, close to the customers	Highly efficient, environmentally compatible processes

Solutions for fuel-efficient cars, energy-efficient houses and renewable energies

Materials that promote health and well-being

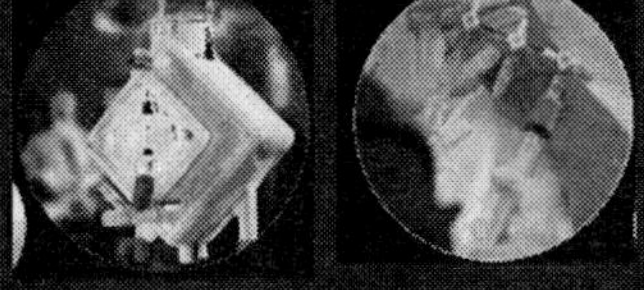

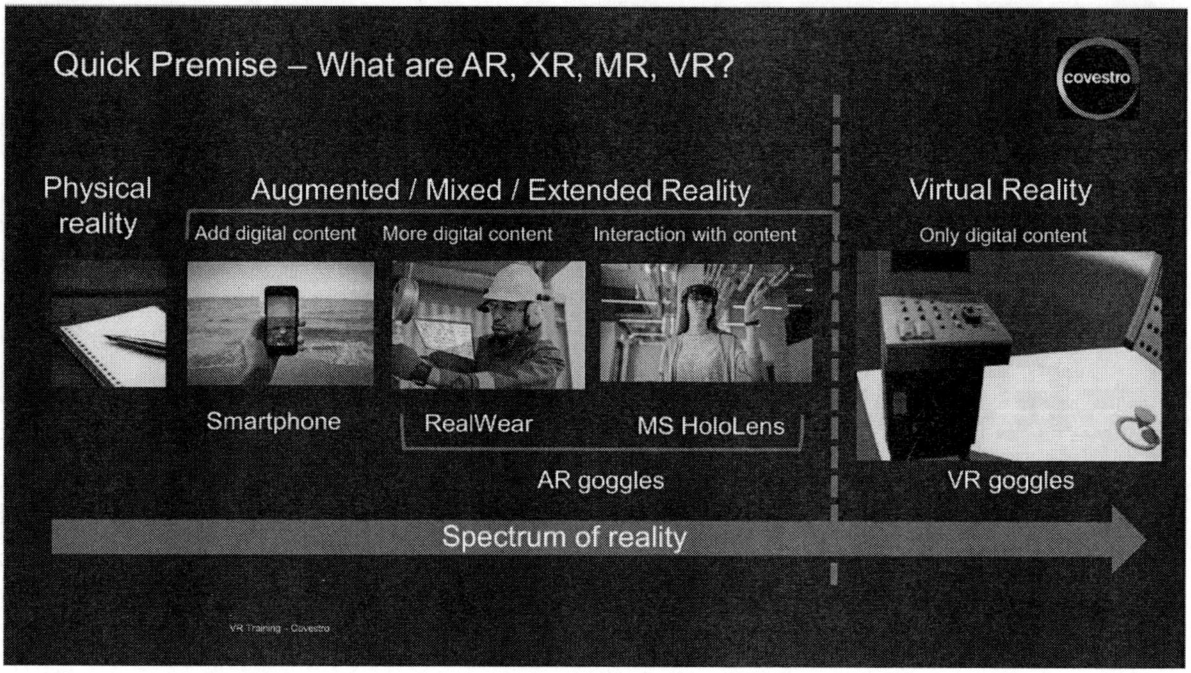
Quick Premise – What are AR, XR, MR, VR?
covestro
Physical reality
Augmented / Mixed / Extended Reality
Add digital content
More digital content
Interaction with content
Virtual Reality
Only digital content
Smartphone
RealWear
MS HoloLens
AR goggles
VR goggles
Spectrum of reality
VR Training - Covestro

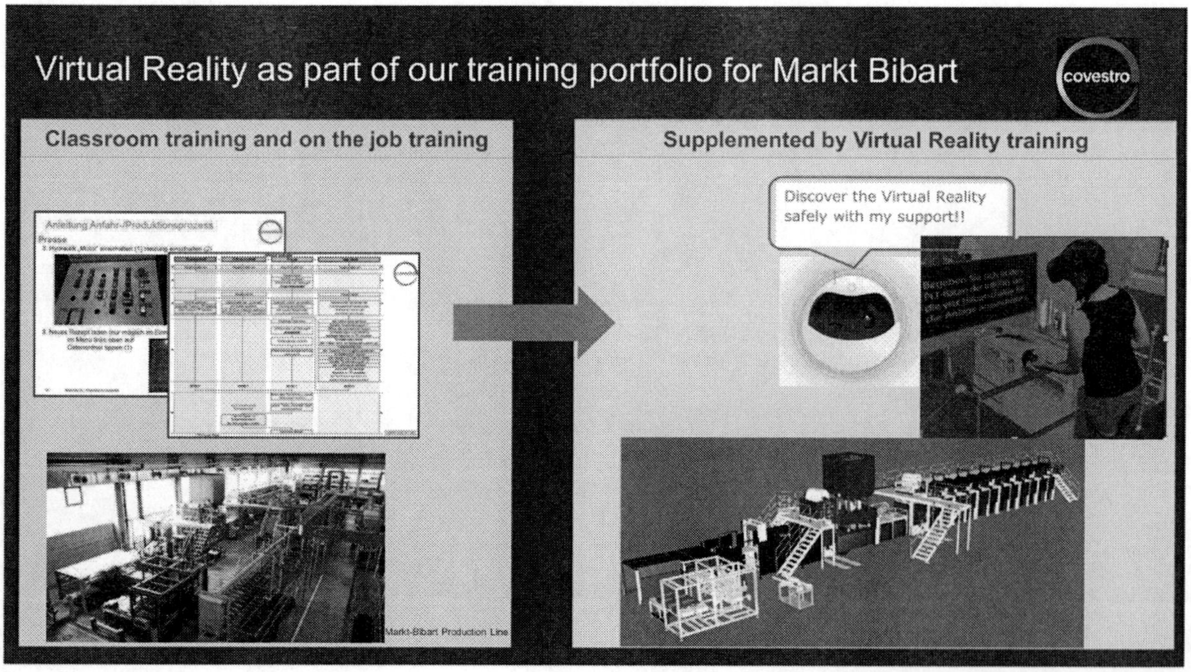
Virtual Reality as part of our training portfolio for Markt Bibart
covestro
Classroom training and on the job training
Supplemented by Virtual Reality training
Discover the Virtual Reality safely with my support!!
Markt-Bibart Production Line

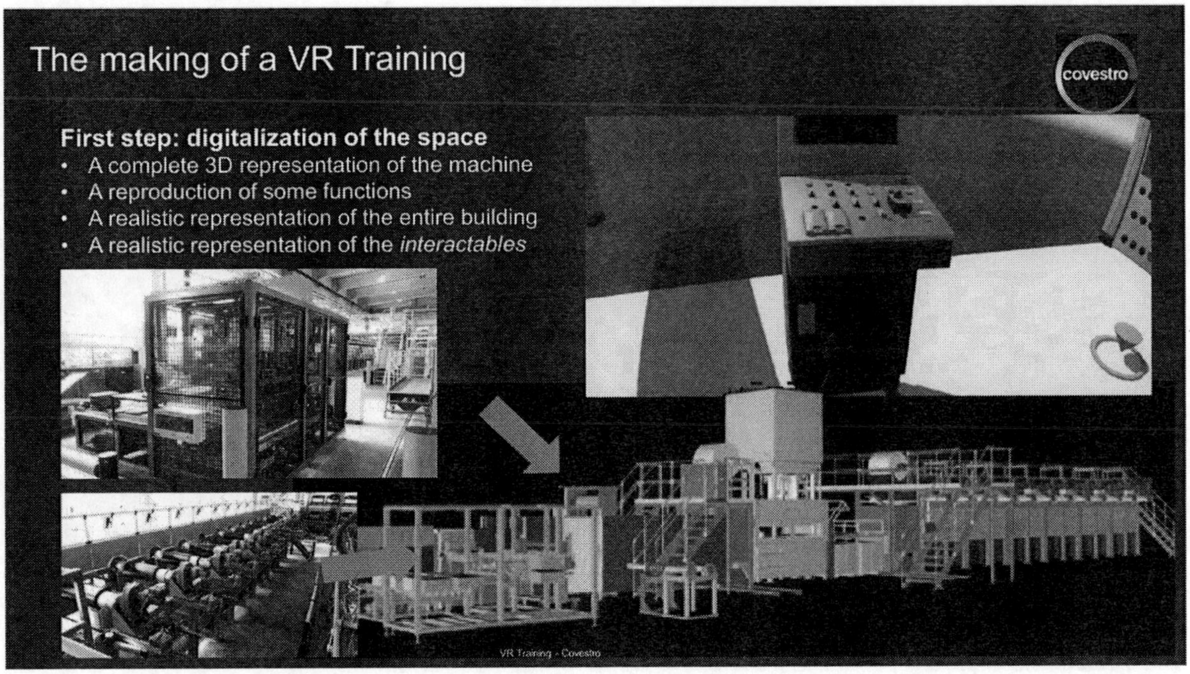
The making of a VR Training
covestro
First step: digitalization of the space
• A complete 3D representation of the machine
• A reproduction of some functions
• A realistic representation of the entire building
• A realistic representation of the *interactables*
VR Training - Covestro

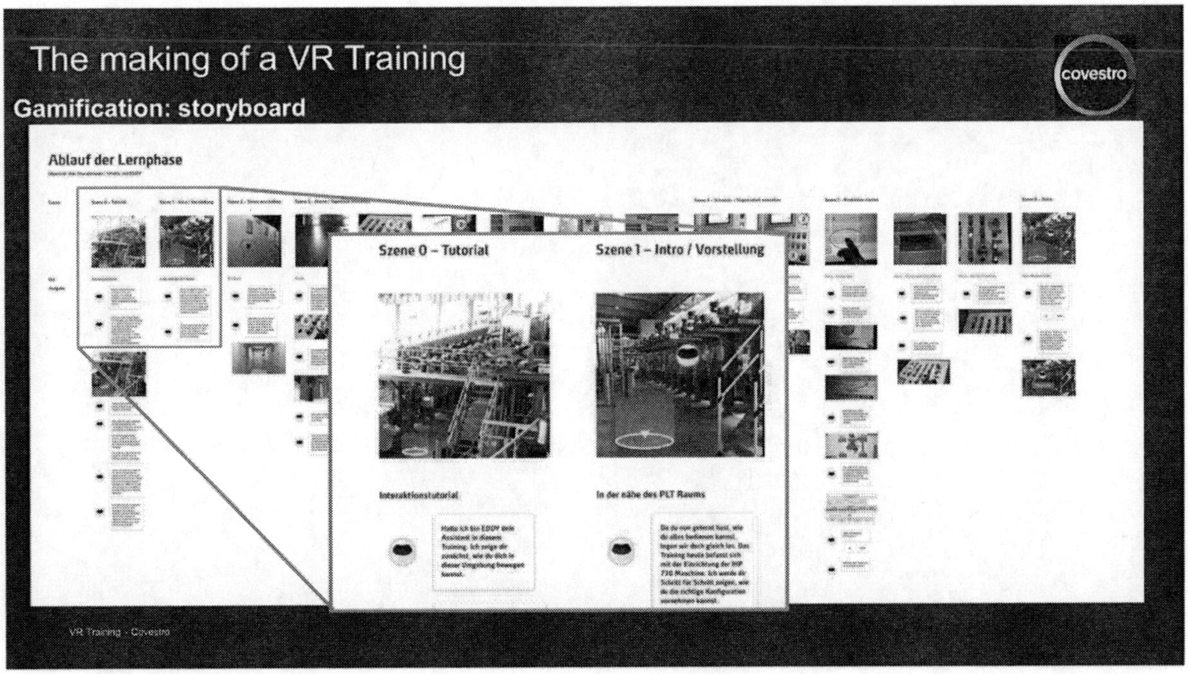
The making of a VR Training
covestro
Gamification: storyboard
Ablauf der Lernphase
Szene 0 – Tutorial
Szene 1 – Intro / Vorstellung
Interaktionstutorial
In der nähe des PLT Raums
VR Training - Covestro

VR for Training – a Pilot in Markt Bibart

Unique opportunity to execute on-the-job trainings in a safe environment

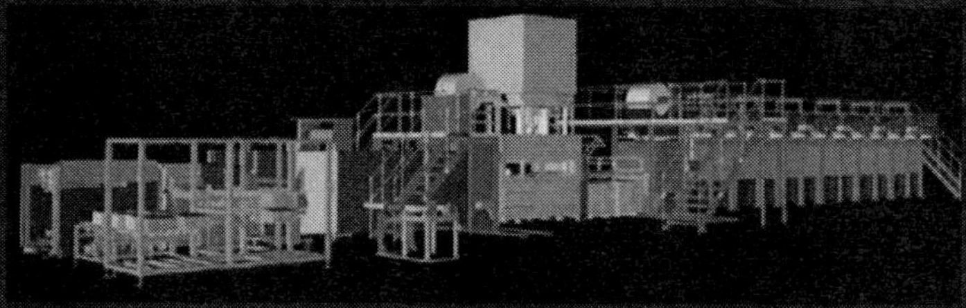

VR Training · Covestro

Advantages of VR Training

Safety

Efficiency

Reliability

Savings

VR Training · Covestro

Thank for your attention.

Questions?

VR Training - Covestro

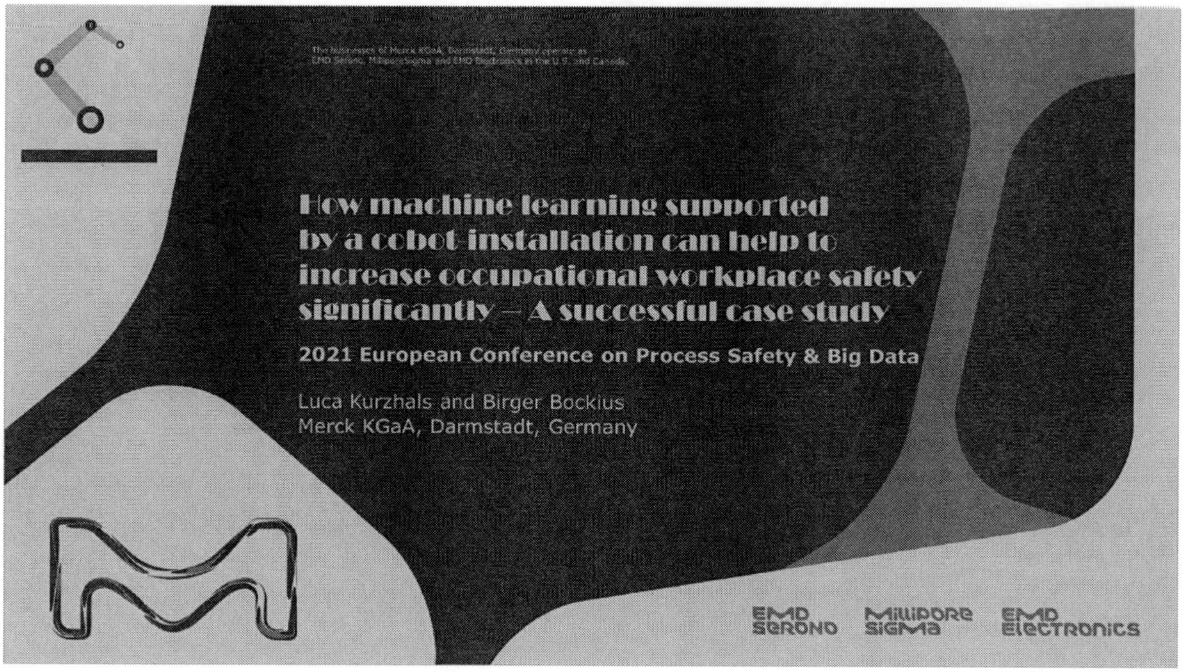
How machine learning supported
by a cobot-installation can help to
increase occupational workplace safety
significantly – A successful case study
2021 European Conference on Process Safety & Big Data
Luca Kurzhals and Birger Bockius
Merck KGaA, Darmstadt, Germany
EMD SERONO
MilliPORE SiGMA
EMD ELECTRONICS

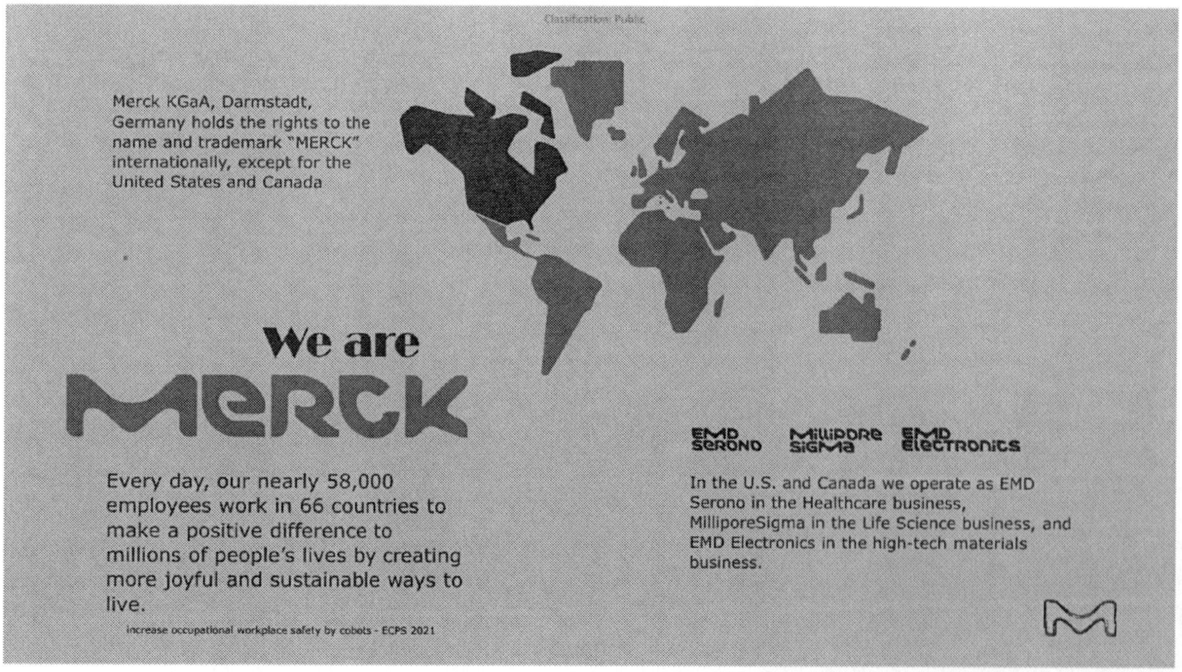
Classification: Public
Merck KGaA, Darmstadt, Germany holds the rights to the name and trademark "MERCK" internationally, except for the United States and Canada
We are
MERCK
Every day, our nearly 58,000 employees work in 66 countries to make a positive difference to millions of people's lives by creating more joyful and sustainable ways to live.
EMD SERONO
MilliPORE SiGMA
EMD ELECTRONICS
In the U.S. and Canada we operate as EMD Serono in the Healthcare business, MilliporeSigma in the Life Science business, and EMD Electronics in the high-tech materials business.
increase occupational workplace safety by cobots - ECPS 2021

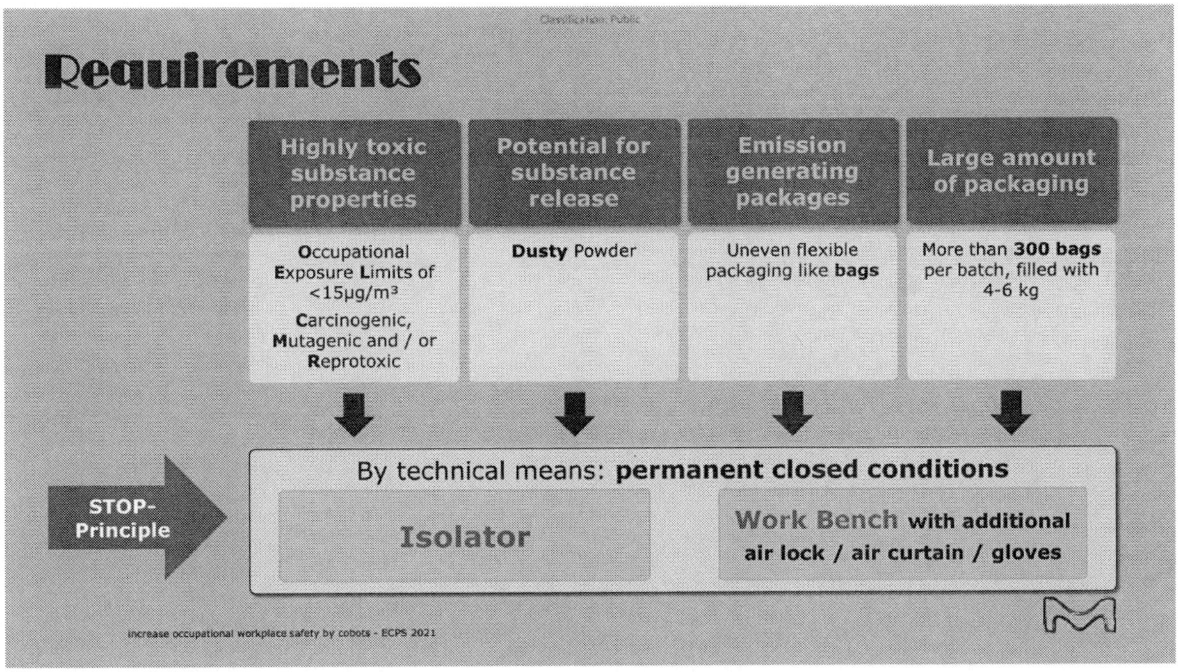
Classification: Public
Requirements
Highly toxic substance properties
Potential for substance release
Emission generating packages
Large amount of packaging
Occupational Exposure Limits of <15µg/m³
Carcinogenic, Mutagenic and / or Reprotoxic
Dusty Powder
Uneven flexible packaging like bags
More than 300 bags per batch, filled with 4-6 kg
STOP-Principle
By technical means: permanent closed conditions
Isolator
Work Bench with additional air lock / air curtain / gloves
increase occupational workplace safety by cobots - ECPS 2021

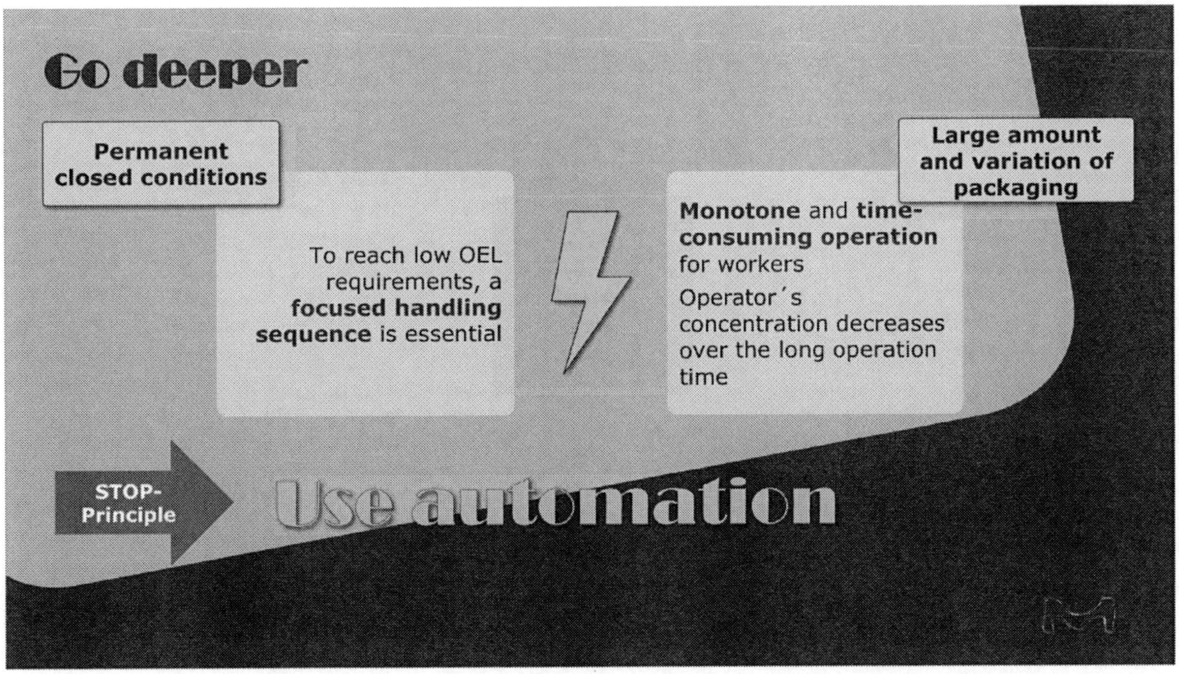
Go deeper
Permanent closed conditions
To reach low OEL requirements, a focused handling sequence is essential
Large amount and variation of packaging
Monotone and time-consuming operation for workers
Operator´s concentration decreases over the long operation time
STOP-Principle
Use automation

Classification: Public

Common Automation Mostly Means...

Additional installation like cylinder, drives, linear units, …
inside containment equipment
will generate **in interaction with powders**
a **significant additional effort** (cleaning, handling with flammable solvents, …)

Really ?

Handling of bags:
Flexible packaging are **not easy to position**
Dimensions and **assembly** of the bags **vary**

cobots - ECPS 2021

Classification: Public

Think new

Think in a new technology concept

Try robotics and a smart optical based machine learning tool
with currently used and optimized equipment

increase occupational workplace safety by cobots - ECPS 2021

Classification: Public

Robot setup

Cobot installation in an adapted isolator

- Enables interaction with operator

+

Closed system (isolator)

- Cobots will be separated from product by gloves (just like the operator)
- Protection against corrosive cleaning solvents

+

Hardware modifications

- To achieve the needed EX-protection
- To create a reliable and safe workflow

ts - ECPS 2021

Classi... ublic

Machine learning (AI) for cobots

- innovative software tool for Cobots: MIRAI by micropsi
- Micropsi industries is supported by M. VENTURES
- revolutionary self-learning is based on camera and sensor-data during recordings (no programming)

How we can use MIRAI:

- Large amounts Force/Torque- and visual data is necessary to handle packaging variations
- Collect data via trainings
- Machine-learning tool to handle these data and create a reliable program
- Enables high flexibility in substance handling

How to train the robot skill:

1. Develop your setup
2. Demonstrate How to React to Variance: Show examples to the robot by hand, how would a human react?
3. Transfer data to Micropsi cloud servers and generate a robot skill
4. Iterate steps 2+3 until you achieve the desired accuracy

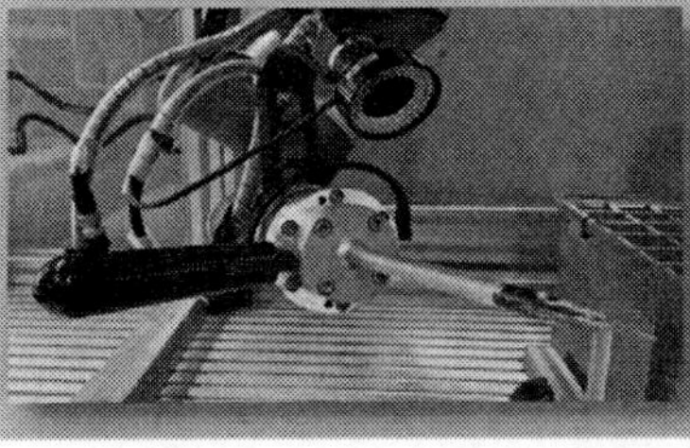

increase occupational workplace safety by cobots - ECPS 2021

Classification: Public

Results of the concept Study

increase occupational workplace safety by cobots - ECPS 2021

Classification: Public

Outstanding challenges

Cobot movement in the isolator is very complex and narrow
→ positioning and motion sequences have to be optimized

Safe and impermeably integration of sensors, camera and gripper onto the gloves
→ has to be integrated into our custom glove-solution

Solvents and hazardous powders resistant gloves are currently not available on the market
→ in cooperation with isolator glove manufactures, we are developing cobot gloves

» Successful case study, more elaboration will be necessary to be ready-for-operate

increase occupational workplace safety by cobots - ECPS 2021

For further discussions or future cooperation, please contact:

Luca Kurzhals
Engineer
EL-OTE-TC
Luca.Kurzhals@merckgroup.com

Birger Bockius
Senior Engineer
EL-OTF-C
birger.bockius@merckgroup.com

VIRTUAL

2021 EUROPEAN CONFERENCE
ON PROCESS SAFETY & BIG DATA

CCPS EPSC

OCTOBER 13-14, 2021

Benefits of Process / Functional Safety Digital Transformation

Mike Scott, CFSE

- Co-chair of ISA S84 committee on Electrical/Electronic/Programmable Electronic Systems (E/E/PES) for Use in Process Safety Applications
- Granted 7-US Patents on Safety Lifecycle
- IEC 61511 committee member
- ISA Fellow
- Certified Functional Safety Expert (CFSE)
- BS. Mechanical Engineering
- Masters of Engineering
- Licensed Professional Engineer

CEO / Co-Chair ISA S84 Committee
Cell +1 (907) 301-3111
mike.scott@aeshield.com

What is Digital Transformation of Process / Functional Safety?

Digitization moving from analog to digital

Digitalization is using digital data to simplify how you work

Digital transformation adds value to every customer interaction

Current Industry Approach

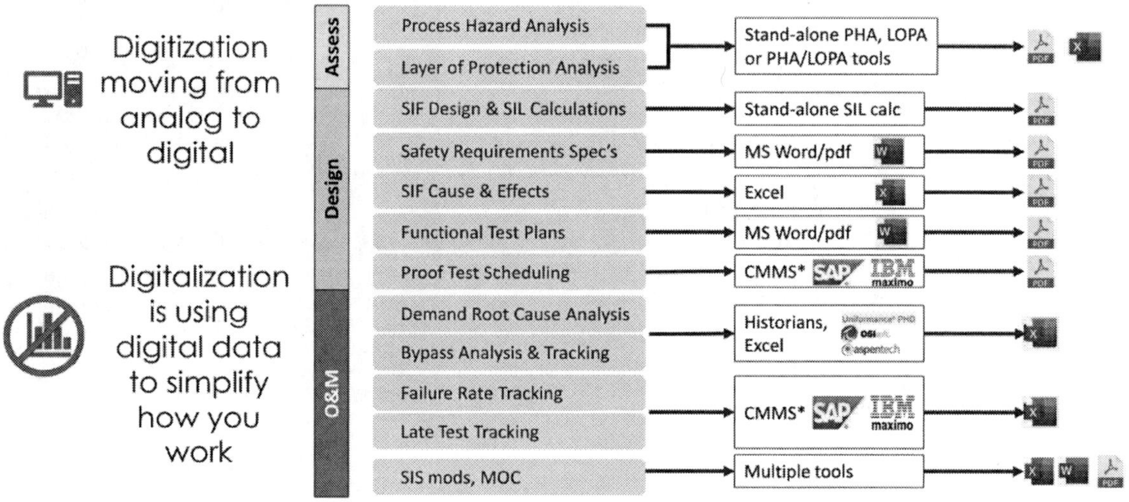

IE 61511 Digital Transformation

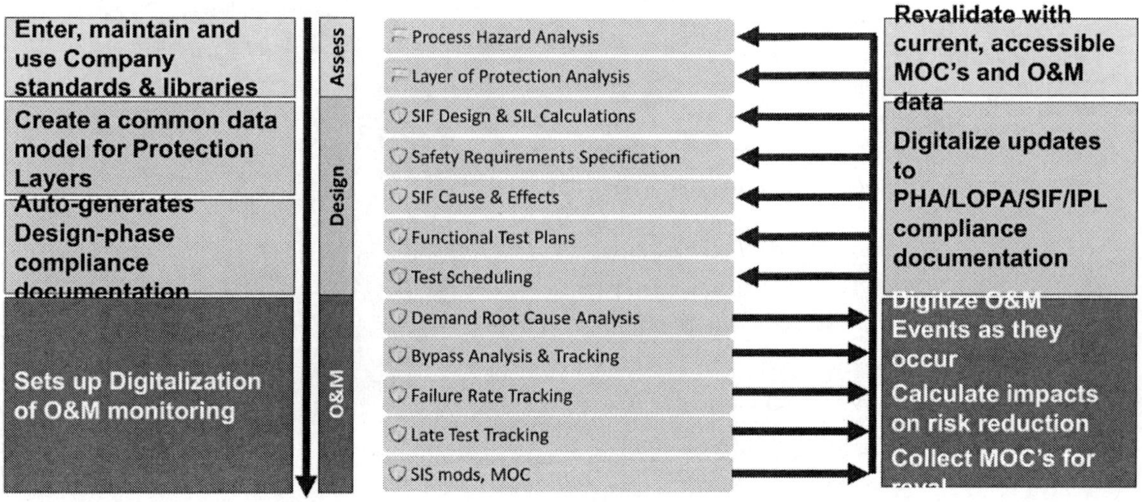

Survey on Benefits of Digital Transformation

- Conducted an exhaustive industry survey in process industry (e.g., Refining, Oil & Gas Production, Bulk & Specialty Chemicals, LNG, Terminals, etc.)
  - 21 global manufacturers
  - 59,000 Safety Instrumented Functions
  - 134,000 Independent Protection Layers

Productivity Savings

Business Owner	Avg Savings (M)*
Capital Projects	$10.1
O&M	$4.8
Test Interval Extension	$2.7
Totals	$17.6

Digital Transformation Pays!

* - Average potential opportunity savings reported by users

Disclaimer

SIEMENS energy

The information contained in this presentation represents the current view of the authors at the time of publication. Process Safety Management is complex, and this presentation cannot embody all possible scenarios or solutions related to compliance. This presentation is for informational purposes only and Siemens Energy makes no warranties.

Meet Your Presenter

SIEMENS energy

Dr. Nancy Faulk is a CCPS Certified process safety engineer, with thirty (30) years in risk assessment, process safety, process engineering, and education. Currently, a Principal Engineer with responsibility for process safety audits, process safety studies (e.g., PSM, Flare, HAZOP, etc.), inspection data management, quality assurance, guidelines development, and training. Dr. Faulk earned her B.S. and Ph.D. from Texas A&M University's chemical engineering department and is API 580 certified.

Page 3 October 2021

Agenda

1

Challenges with PRV Performance

2

Meeting Design and Inspection Requirements

3

Data Sources

4

Leveraging Data Sources to Address Performance Issues

Challenges with PRV Performance

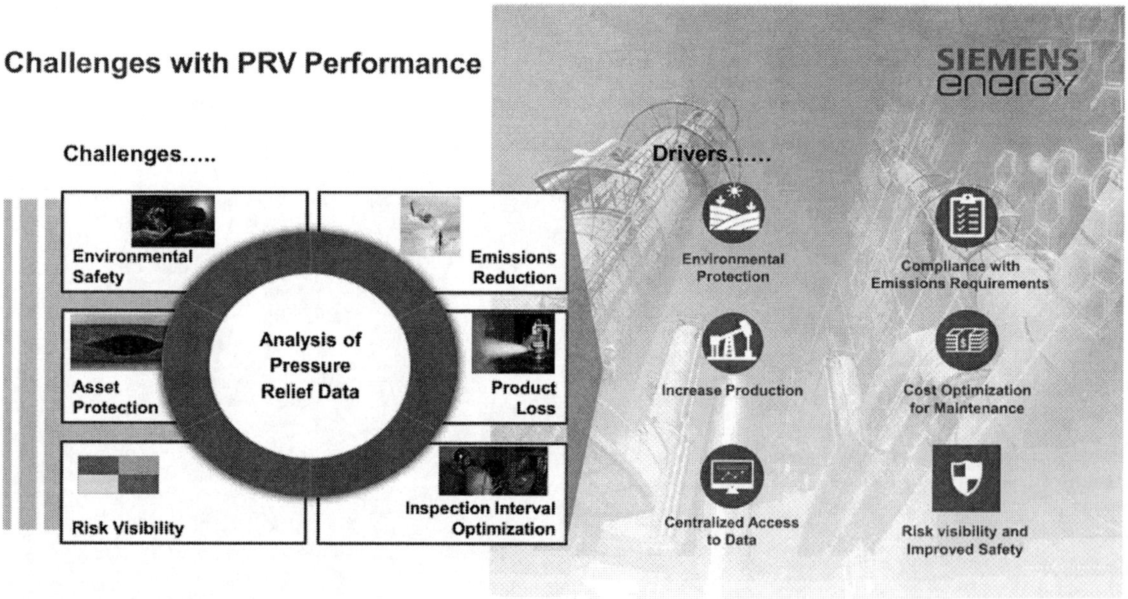

Challenges with PRV Performance

SIEMENS energy

- When maintained properly, a relief device can stay in service 30 years
- PRVs may experience failure by:
  - Failing to open on demand Valve has no pressure
  - Valve is simmering, leaking or **chattering**
- Chattering is high-frequency and high-amplitude movement of opening and reclosing the valve
- Causes include:
  - Pressure drop issues
  - Oversized relief device
  - Relief device which must handle widely varying flow rates

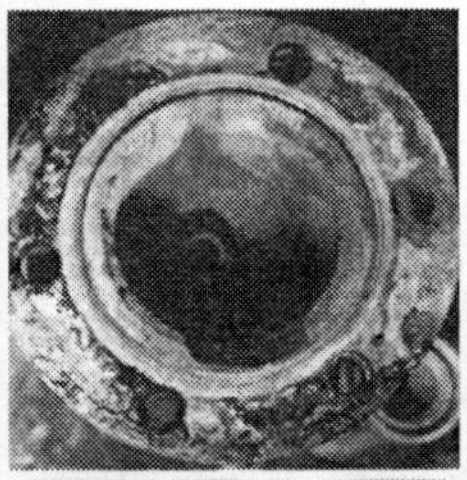

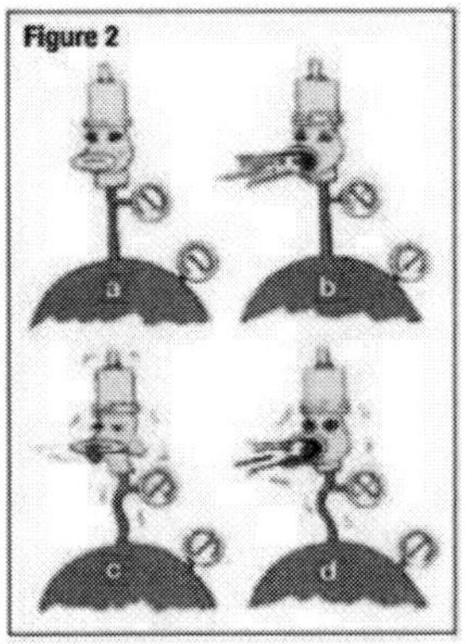

Source: CCPS Process Safety Beacon, April 2013

Leaking Relief Valves

SIEMENS energy

"Even a small leakage (0.1% from the PRV flow area) can cause losses of tens of thousands of dollars per year.
Additionally, the leakage can cause significant emissions violations, resulting in expensive fines and even required shutdowns."

Gas type	Gas per metric ton (USD)[1]	Process pressure (psig)[2]	Leakage yearly losses (USD)
Ethylene	1,044	250 @ 100°C	740,000
Ammonia	500	250 @ -33°C	335,000
Steam	22	250 @ 200°C	7,800

1. July, 2015 Platts Global Petrochemical Prices.
2. Relief valve set pressure – 300 psig and ASME orifice type "G".

https://www.plantservices.com/articles/2009/206/

Source: https://www.processonline.com.au/content/process-control-systems/article/pressure-relief-device-monitoring-how-to-detect-releases-leaking-and-fugitive-emissions-part-1-857338063

Agenda

1

Challenges with PRV Performance

2

Meeting Design and Inspection Requirements

3

Data Sources

4

Leveraging Data Sources to Address Performance Issues

Meeting Design and Inspection Requirements

SIEMENS energy

- PRVs are relatively simple and reliable
- The simplicity of mechanical operation is overshadowed by:
  - Design requirements
  - Inlet and outlet pressure drop allowed
  - Sizing and selection
  - Installation
  - Inspection and maintenance requirements
- Requires a full appreciation of non-design basis

Meeting Design and Inspection Requirements

SIEMENS energy

- Normal interval between shop inspection/overhaul typically:
  - 5 years for process services
  - 10 years for clean services
- Factors which may shorten interval include:
  - corrosive or fouling service
  - **subject to vibration**
  - frequent operational upsets
  - **leakage problems**

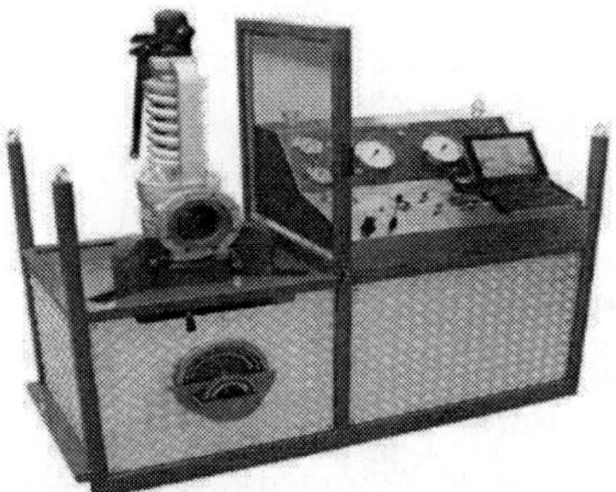

http://www.directindustry.com/prod/venti-test-equipment-bv/product-61466-1071795.html

We need to know <u>before</u> a bench test is scheduled if:
- **A valve is being damaged by vibration**
- **Leakage and loss of product is occurring**

Agenda

1

Challenges with PRV Performance

2

Meeting Design and Inspection Requirements

3

Data Sources

4

Leveraging Data Sources to Address Performance Issues

Data Sources

SIEMENS energy

1 Sources

Industrial Process

2 Software Programs, Modeling, & Simulations

3 Manual Inputs

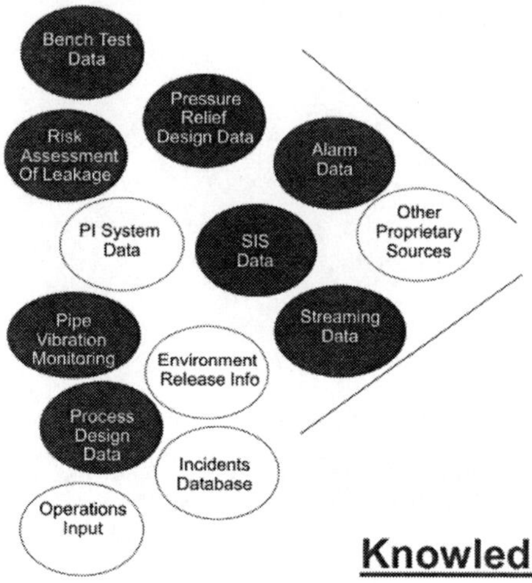

Actionable Insight

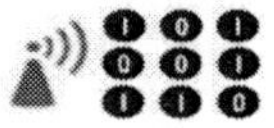

Generating big data

Knowledge/Intelligence

Page 13 October 2021

Agenda

1

Challenges with PRV Performance

2

Meeting Design and Inspection Requirements

3

Data Sources

4

Leveraging Data Sources to Address Performance Issues

Page 14 October 2021

Leveraging Data Sources to Address Performance Issues

SIEMENS energy

Data analytics algorithm

- Verify adequacy of relief device
- Detect leakage
- Assess stability of relief device
- Optimize inspection intervals

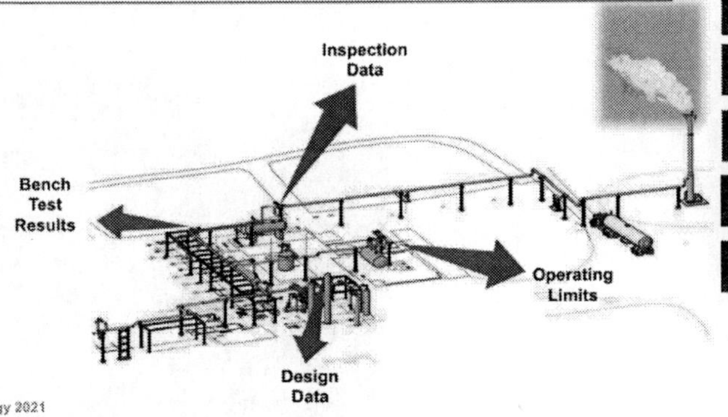

- Reduce emissions
- Ensure safe operations
- Environmental & safety regulatory compliance
- Operations optimization for maximum safe throughput
- Effective decisions on predictive & preventive maintenance

Relevance to Asset Lifecycle

Improved risk visibility through data digitization

SIEMENS energy

Centralized Data Management System → **Data Analytics & Visualization** →

Leading Indicators
Lagging Indicators

Benefits of this data analysis ...

- ... can help users **understand** which devices are **emissions sources** in installed pressure relief systems and use that information to reduce hydrocarbon release
- ... can be used to **plan budgets and resources** for optimized PRV maintenance
- ... can be used to **prioritize environmental safety** by preventing leaks and losses
- ... can improve **flaring optimization and reduce venting** of hydrocarbons, representing lost product and revenue

Summary

- Have seen the challenges and drivers for reducing emissions from relief devices
- These same drivers also help to protect the relief device from valve damage, as well as prevent the financial cost of lost product, valve repair, buying carbon credits
- The proposition is to leverage big data as a tool to reduce leakage and ensure safe operations

Please let us know how we can support your business needs!

SIEMENS energy

Siemens Process & Safety Consulting

15375 Memorial Dr, Suite 700
Houston, TX 77079, USA
Phone: (713) 570-2900
Phone: +1 (800) 658-8809
Fax: (713) 570-2999

info.ogconsulting.energy@siemens.com

siemens-energy.com/process-safety

Aggregating OT data to promote operability and safety
European Conference on Process Safety and Big Data 2021
Dr. Sven Lohmann
EMERSON
Emerson Confidential

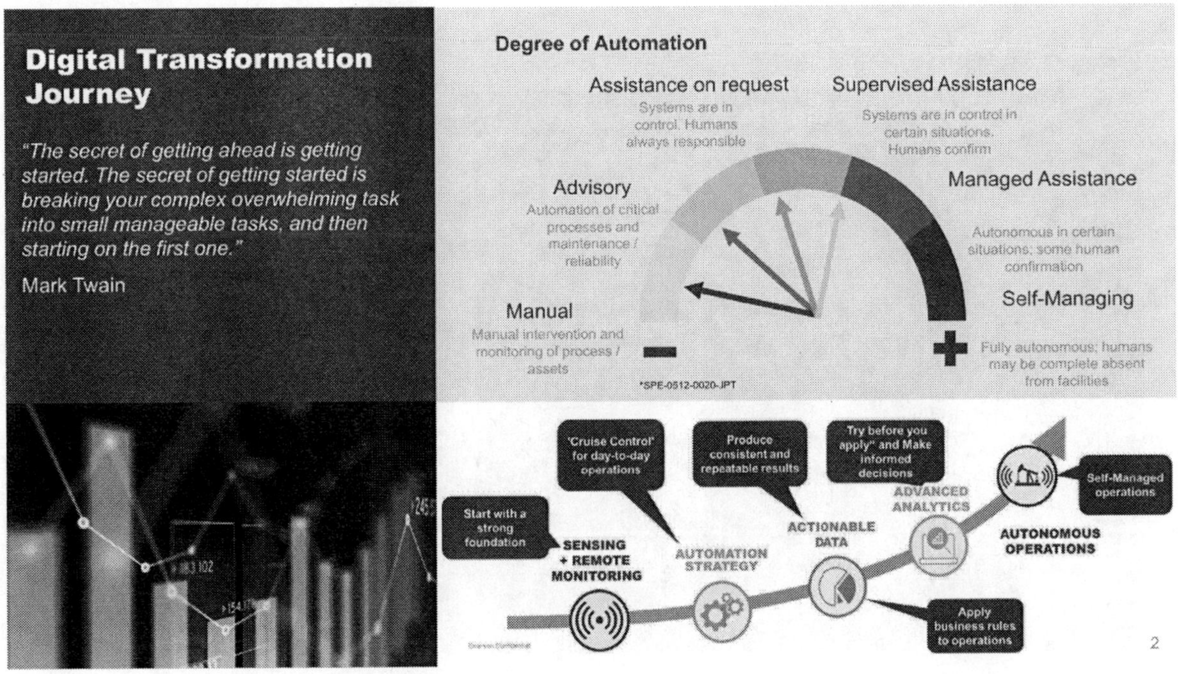
Digital Transformation Journey
"The secret of getting ahead is getting started. The secret of getting started is breaking your complex overwhelming task into small manageable tasks, and then starting on the first one."
Mark Twain
Degree of Automation
Assistance on request
Systems are in control. Humans always responsible
Supervised Assistance
Systems are in control in certain situations. Humans confirm
Advisory
Automation of critical processes and maintenance / reliability
Managed Assistance
Autonomous in certain situations; some human confirmation
Manual
Manual intervention and monitoring of process / assets
Self-Managing
Fully autonomous; humans may be complete absent from facilities
*SPE-0512-0020-JPT
'Cruise Control' for day-to-day operations
Produce consistent and repeatable results
Try before you apply" and Make informed decisions
Start with a strong foundation
SENSING + REMOTE MONITORING
AUTOMATION STRATEGY
ACTIONABLE DATA
ADVANCED ANALYTICS
AUTONOMOUS OPERATIONS
Self-Managed operations
Apply business rules to operations
2

SAFETY

Zero injuries. Zero incidents.

- Employees removed from hazardous activities
- Constant monitoring of worker and plant health
- Threat prevention and response – cyber, physical – is automated, part of culture

RELIABILITY

No unplanned downtime at minimal cost

- Analytics predict health of all equipment
- No reactive repair: pre-failure corrective action by automated closed-loop systems
- 'Avatars' perform field tasks

THE GOALS OF DIGITAL TRANSFORMATION FOR OPERATIONS

PRODUCTION

Production optimized to market conditions

- Continuous visibility enables active business management
- Business areas optimized in real-time across interconnected systems
- Predictive digital twins enable exception management along entire value chain

ENERGY & EMISSIONS

Recognized sustainability leader

- Continuous analytics predict and prevent releases
- Energy consumption measured and dynamically optimized with production
- Energy sources balance business and shared community goals

WORKFORCE

Empowered workers drive even more value

- All routine tasks automated: staff takes larger role in meeting KPIs
- Collaboration embedded in culture and work processes
- Faster, better decisions supported by analytics and on demand expertise
- Career-long development and certification through virtual tools

3

USE-CASES

Smart. Connected. Living. Learning.

VISUALISATION & ALERTS

View performance from C-suite to plant floor across all areas of plant
Integrated alerting for the right people at the right time from any source

ADVANCED ANALYTICS

Predictive insights using advanced analytics moving from reactive to proactive
Link analytics to dynamic plant scheduling tools to enable agile operations

ACTIVE PLANT & BEST NEXT ACTION

A digital representation of the plant combining real-time data and analytics
Artificial Intelligence integrated with plant data to guide next best action

PREDICTIVE INTEGRATED OPTIMISATION

Optimise plant performance beyond functional silos by using integrated advanced analytics models

AUTONOMOUS OPERATIONS

Reduce risk and cost through autonomous operations using Artifical Intelligence techniques such as deep learning

DIGITAL WORKFORCE

Utilize automated workflow driven mobile solutions such as ATEX certified tablets, smart phones, Industrial glasses with augmented and virtual reality to overlay real time operational data, work history, asset conditions for operations and maintenance workers. Reduce Risk, Improve safety and optimize costs through self directed workforce.

4

Where does OT data come from?

How do you identify problems?

- Manual Input (File)
- Semi-manual feedback
- Motors
- Turbines
- Instruments
- Control Loops
- Distillation Columns
- Heat Exchangers
- Historian-Servers
- PLCs

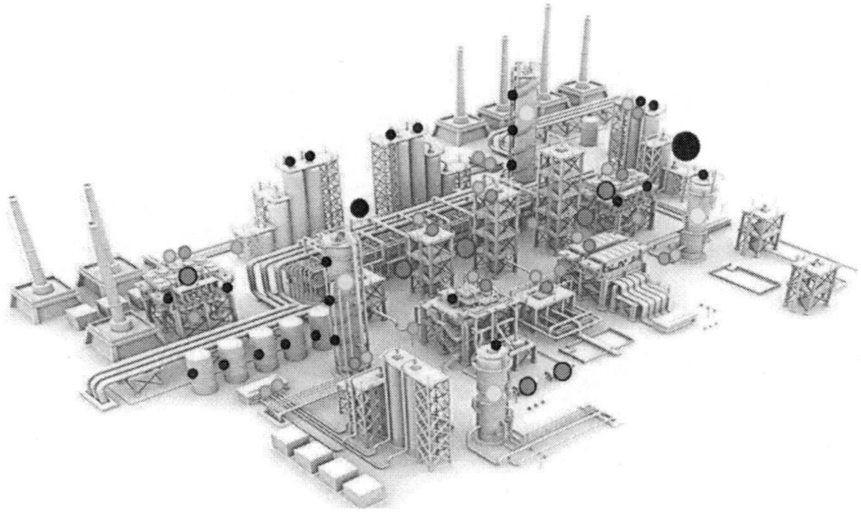

Emerson Confidential

5

Production Sites Are not Self-Sufficient Anymore ...

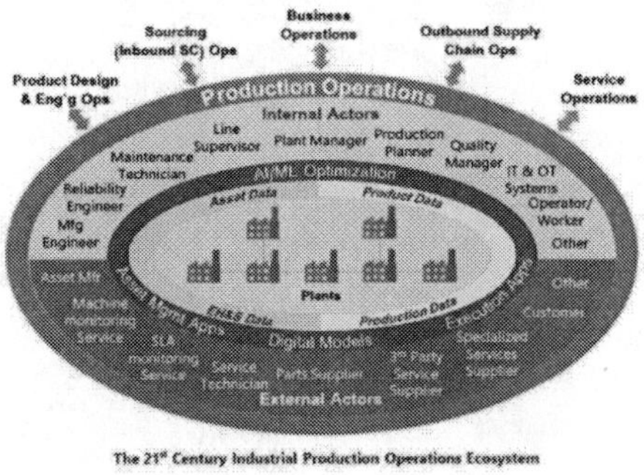

The 21st Century Industrial Production Operations Ecosystem

Source: https://www.arcweb.com/blog/framework-digital-transformation

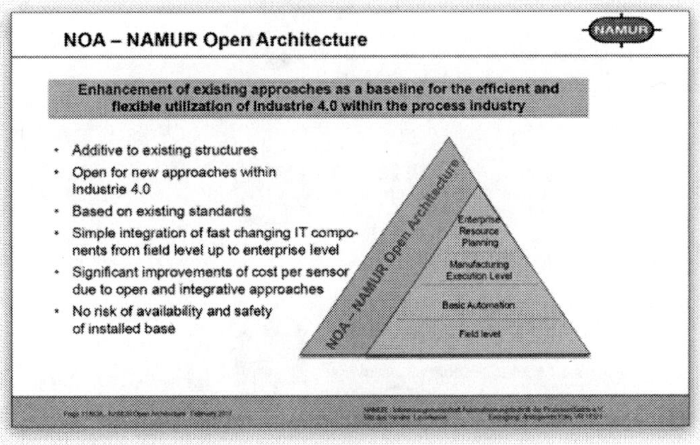

6

Wish List - Characteristics

- Global fleet updates executed in background
- Designed for multinational operators

- Easy and flexible
- Federated or centralized
- Centrally managed

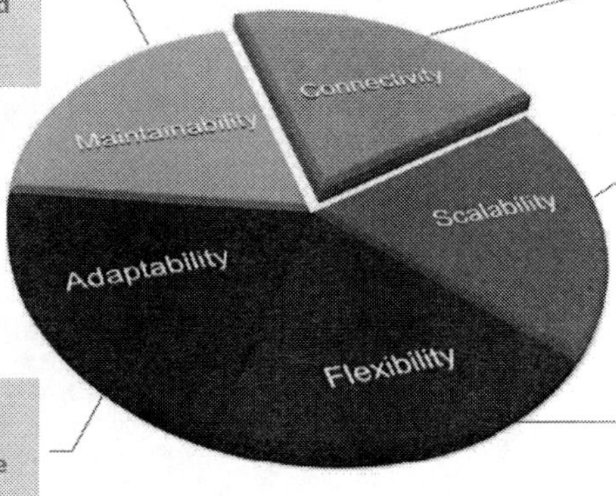

- Small to very large systems
- Storage can be on-prem, cloud based or mixed

- Adapts to various use cases
- Allows to harmonize the data of underlying systems on-the-fly

- Highly customizable
- Fully open

7

Wish List – Data Sources

SENSORS
A limitless variety of custom sensors has been developed to gather data on any relevant information such as geolocation, temperature, vibration, biometrics, ...

ASSET DATA
Modern industrial equipment provide a stream of real time data to support optimization analysis.

TRANSACTIONAL DATA
Any transaction handled by the ERP leaves a detailed record with relevant information.

INTERNET & MOBILE DATA
Social media generated data, news, published statistics and other internet available sources can be analyzed and interpreted.

IMAGES
Both static and dynamic images are now all digitalized and interpretable by complex photo/video analytics techniques.

TEXT
Optical character recognition is now capable to transform physical and electronic text into data.

TRAPPED VALUE

"Most companies are collecting data for the sake of collecting data, and struggle to "unlock" the Trapped Value

8

Data Lakes | Security: Data Governance

- Eco-system designed for high security and firewalls/authentication.
- Security data models for additional layers of data egress security.
- Network Connection level through TLS (Transport Layer Security) protocol (IETF RFC 5246) together with Secure Remote Password (SRP) authentication (IETF RFC 5054).
- Certificate authentication via X.509 CA standard.
- Data Repository (MongoDB) can be hosted on-prem or natively on the cloud. Supports IETF Standard RFC 5802.
- Semantics handles & configured at one level.
- Relay service available to provide firewall friendly data transfer from L2/L3 to L4

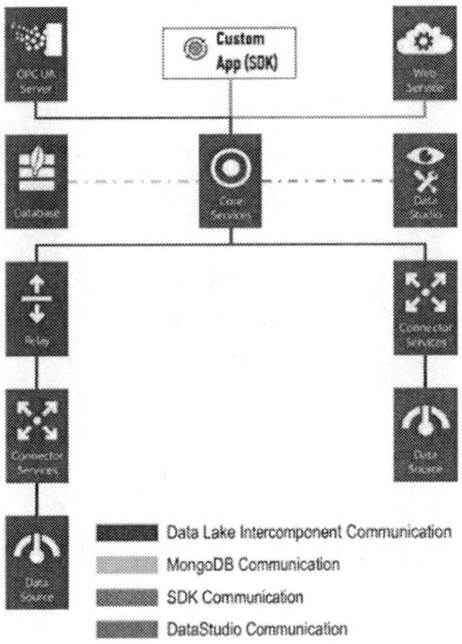

Each site can host a "Local Core" that can connect to a "Global Core" on the enterprise level that, although being part of the overall system, can run full self-sufficient.

9

IIOT architecture - Unified Namespace

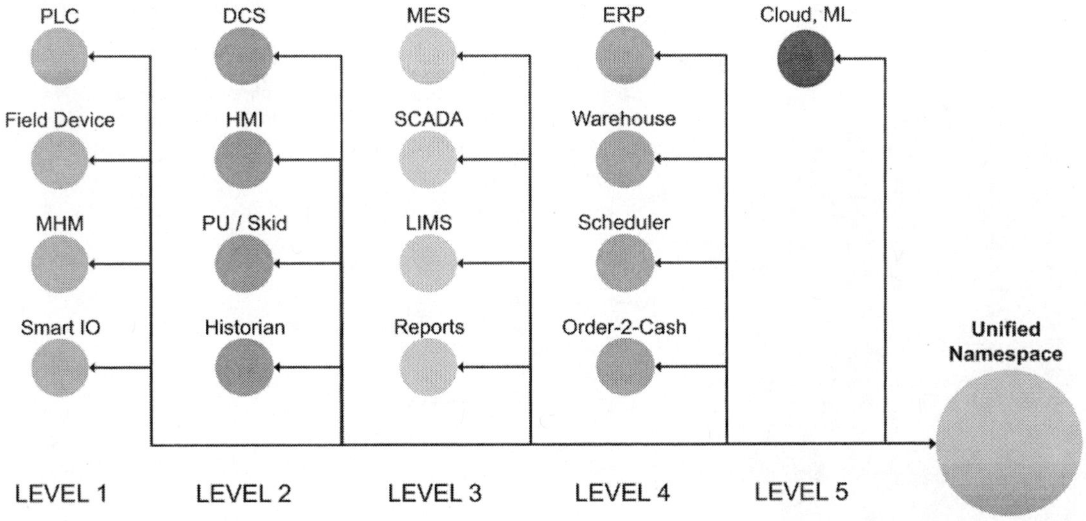

10

IIOT architecture – grown reality in many cases

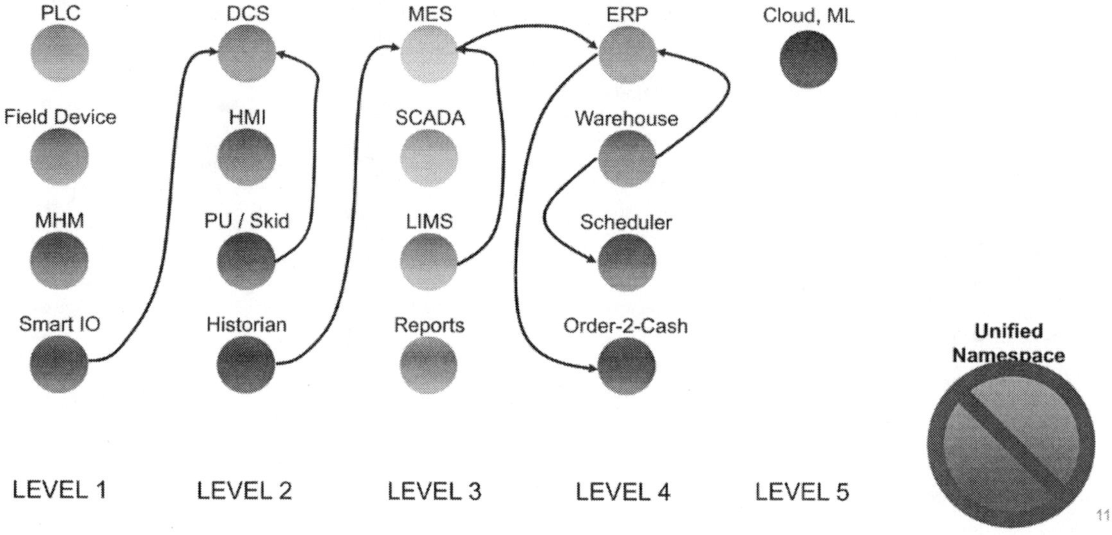

11

Lifecycle considerations - House of Cards

- Over the lifecycle the point-2-point connections of the corporate-wide data integration resembles a **house of cards**
- Even if carefully built – the house of cards does not cope easily with changes
- Several changes put the house at risk of collapse

Central management is key for success – structurally & conceptually

Emerson confidential

12

Plantweb Optics Data Lake – Service-based architecture

- Data Source is connected to the Core through the Connector. This could be an OPC data source, a relational database, or other custom options.
- Connector connects to different types of data sources and forwards the data to the central Core service.
- Core is the central component of the system. It establishes connections to multiple Connector services to collect data from various data sources. The Core processes the data, storing all information in the Data Repository.
- Data Repository is a data store using mongoDB, a free, open-source NoSQL database. mongoDB is infinitely scalable to deal with vast amounts of real-time data.
- Interface Servers expose data in the Plantweb Optics Data Lake according to the latest OPC and WebAPI specifications.

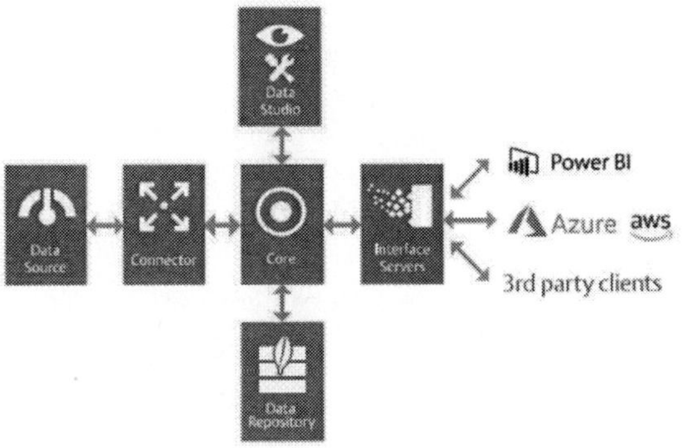

Examples: Safety

Collect HSE data

- Record of "near misses" in the field
- Results of interviews with people on site
- Collect data comfortably from an Excel-spreadsheet or website

Increase acceptance by using the tools that are accepted

Automated Diagnostic Data

- Track bridged safety loops with real-time data
- Visualize those that have been active more than 24 hours
- Show those that do NOT have a timed latch on them

Create visibility and thus raise awareness

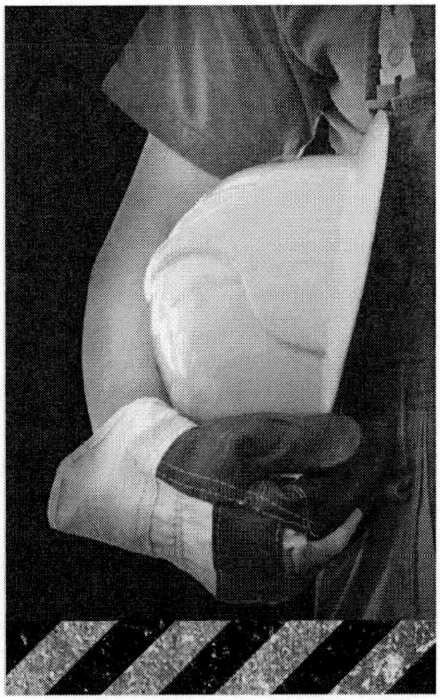

Examples: Operability

Increased Awareness

- Make data available where it is needed – e.g. the mobile of the responsible worker/engineer
- *Actionable data* increases awareness and reduces the delay time that may occur

Channel the info where it belongs and where it is best responded to

Prescriptive Maintenance

- Analytics employing first principal models and/or operational data and machine learning allow forward planning
- Automated process insights enable automated results that are prescriptive in nature – thus reducing response times

Employ model-based analytics tailored for purpose

2021 EUROPEAN CONFERENCE ON PROCESS SAFETY & BIG DATA
October 13–14, 2021 CCPS EPSC

VIRTUAL

PANEL 2

The journey for sustainable and innovative implementation of process safety big data analytics projects

Panel Moderators

Dr. Abdul Aldeeb
Siemens Energy

Tony Downes
Honeywell

Co-organized by the Center for Chemical Process Safety and the European Process Safety Centre

2021 EUROPEAN CONFERENCE ON PROCESS SAFETY & BIG DATA
October 13–14, 2021 CCPS EPSC

VIRTUAL

The journey for sustainable and innovative implementation of process safety big data analytics projects

Panelists

Emmanuel Romeu, SAS (JMP), Sr Systems Engineer

Dag Yemenu, ISN NetWorld, Executive Vice President of Products

Joe Stough, SIFZero Advisory Services, Founder

Co-organized by the Center for Chemical Process Safety and the European Process Safety Centre

2021 EUROPEAN CONFERENCE ON PROCESS SAFETY & BIG DATA

October 13–14, 2021

CCPS EPSC

VIRTUAL

Emmanuel Romeu is JMP Sr Systems Engineer at SAS.

Prior to joining SAS, he spent 25+ years within Beckman-Coulter developing diagnostic tools for fertility, bone and infectious diseases. During his last ten years he have developed and expanded his statistical expertise to accelerate time to market, improve product quality, safety, manufacturing reliability and facilitate complex regulatory submissions to the FDA, cFDA, and European Commission.

Emmanuel is passionate about Statistics and its spread across organizations.

Co-organized by the Center for Chemical Process Safety and the European Process Safety Centre

2021 EUROPEAN CONFERENCE ON PROCESS SAFETY & BIG DATA

October 13–14, 2021

CCPS EPSC

VIRTUAL

Dag Yemenu is Executive Vice President of Products at ISN. In his role, Dag is responsible for leading the strategy and execution aspects of the company's products. Prior to his current role, he held various leadership positions in the company's operational groups in the U.S. and abroad.

Before joining ISN, Dag had engineering careers in the Power Generation and Aviation industries. He holds a Bachelor's degree in Mechanical Engineering, a Master of Science degree in Industrial Engineering and an MBA.

Co-organized by the Center for Chemical Process Safety and the European Process Safety Centre

2021 EUROPEAN CONFERENCE ON PROCESS SAFETY & BIG DATA

October 13–14, 2021

CCPS EPSC

VIRTUAL

Joe Stough is the Founder of SIFZero Advisory Services. Joe has over 25 years founding and running enterprise software and data analytics solutions in the Safety and Operational Excellence market. Recently founded SIFZero to deliver advisory services in same space.

Co-organized by the Center for Chemical Process Safety and the European Process Safety Centre

Emmanuel Romeu
Sr Systems Engineer - Chemistry Europe
emmanuel.romeu@jmp.com

Software solutions with sophisticated data analytical capabilities

jmp STATISTICAL DISCOVERY FROM SAS

ISN Collect. Verify. Connect.

Supply Chain Sustainability

Insights from Contractor Safety Incident Analysis

October 2021

www.isn.com

ISNetworld is an **online contractor and supplier management platform** of data-driven products and services that **help manage risk, support sustainability and strengthen relationships.**

650+
Hiring Clients Worldwide

75,000+
Active Contractors/Suppliers

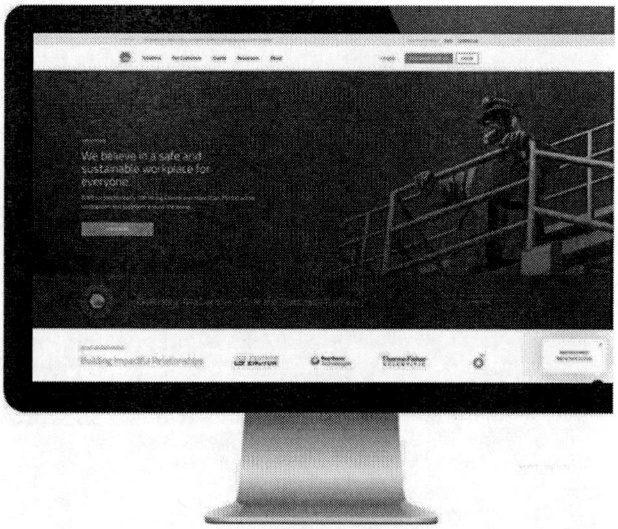

Outsourcing Outlook and Top Priorities for Hiring Organizations

95%

of organizations believe outsourcing work to third-party contractors will increase or remain at the same over the next 12-24 months.

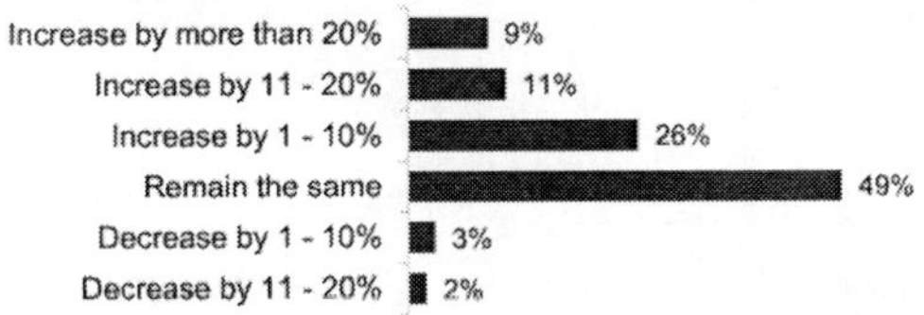

*SIF – Serious Injuries and Fatalities
**What are the major priorities that will drive your organization's contractor management objectives in the next 1-2 years?
Survey data n = 233 decision makers from 150 organizations.

Top Priorities for Hiring Organizations**

- Safety Culture Alignment with Contractors
- Improving the safety performance of contractors (including SIF*s)
- Improving Process Safety / Preventive Maintenance Management

Serious Incidents and Fatalities** (SIFs) Analysis Lifecycle

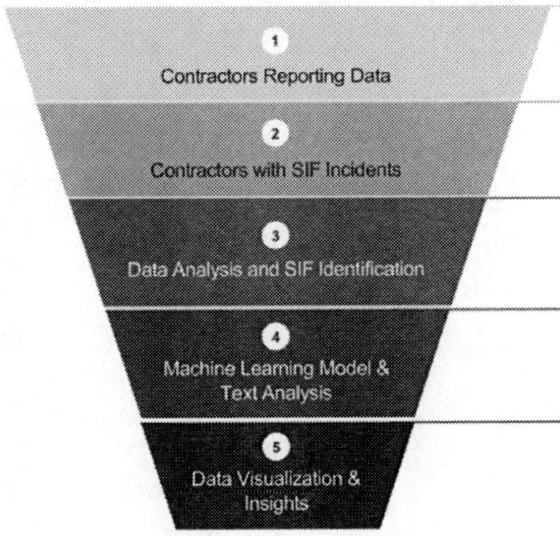

50,000+ US Contractor companies

2,600+ Contractors with SIF incident(s)
7,500+ OSHA logs containing **60,000+** incidents

18,000+ SIFs identified
Serious Injuries and Fatalities cases were classified

A deep learning model (BERT) used to classify incident details
Model was trained on OSHA severe injury dataset*

Tableau dashboards for interactive insights

*https://www.osha.gov/severeinjury
** 2018, 2019, 2020

Serious Injuries and Fatalities (SIFs) Analysis – Petrochemical* Industry

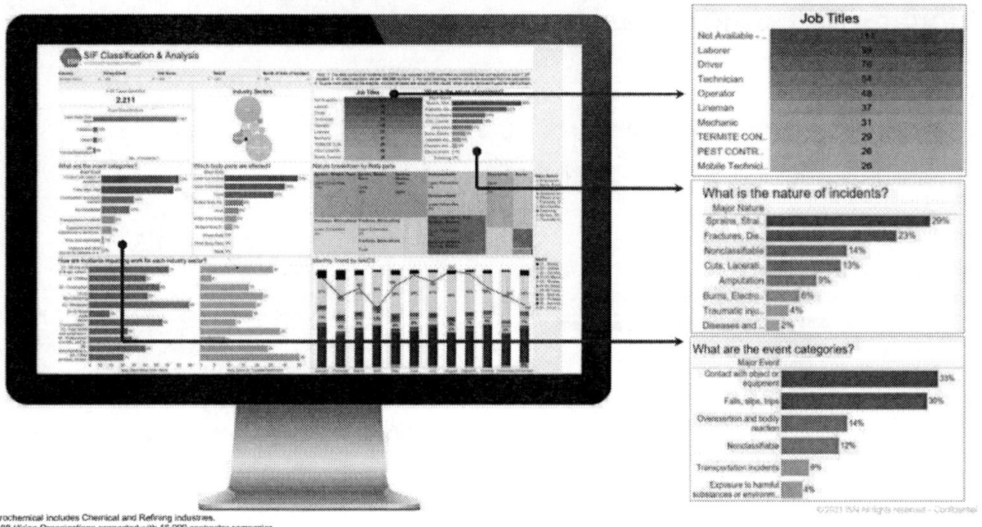

*Petrochemical includes Chemical and Refining industries.
n =108 Hiring Organizations connected with 15,000 contractor companies.

Serious Injuries and Fatalities (SIFs) Analysis – A Hiring Organization's View

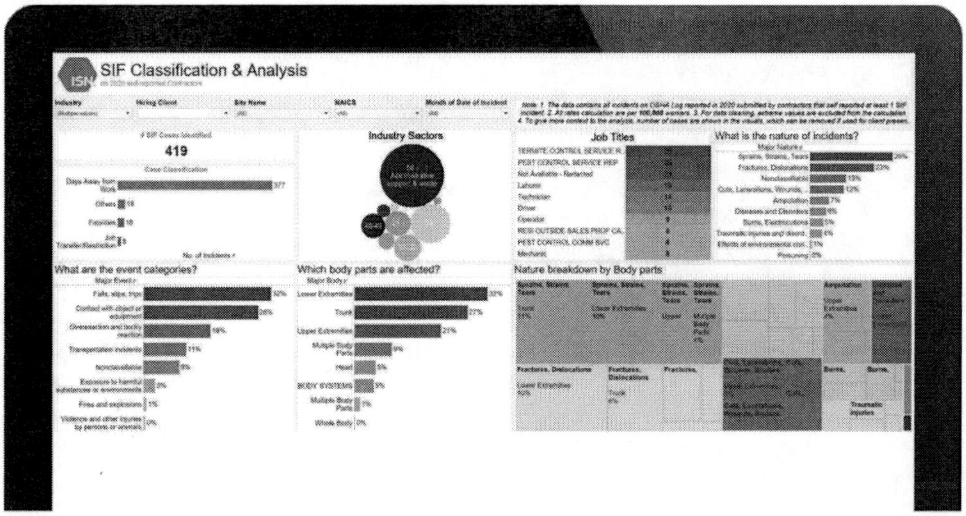

Serious Injuries and Fatalities (SIFs) Analysis – A Hiring Organization's View

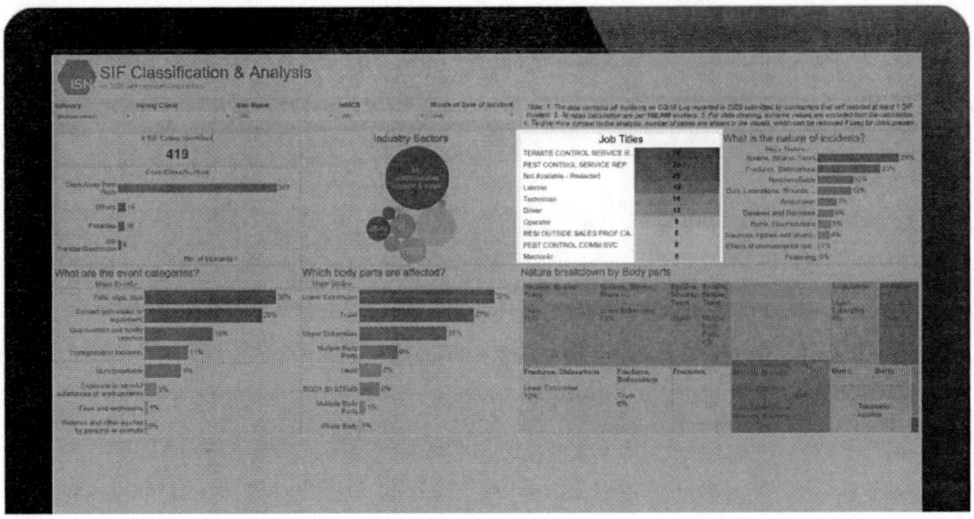

Creating Ongoing Value to Stakeholders

1. Self-service portals and dashboards empower end users
2. Trending and benchmarking adds perspective to data
3. Predictive analytics provides foresight

Analytics of Safeguard Improvement Capability to Reduce Incidents

Joe Stough
October 14, 2021

Start with End in Mind ➔ Zero Incidents

- Weak safeguards (Physical, Organizational) are cause of <u>every</u> incident
- Perfect Safeguards ➔ ZERO Incidents
- Stronger Safeguards ➔ Fewer Incidents
- Hypothesis behind a decade of Predictive Analytics:
  - Stronger Safeguard Improvement Capability (X's) ➔ Fewer Incidents (Y)
  - Index of actionable factors ➔ **Engagement + Discipline + Materiality-driven**

Joe Stough – Career of Software, Data & Analytics

Founded Safety & OE Enterprise Software Co.	Deployed Worldwide > 100 Countries	Safety and OE Predictive Analytics & Benchmarking	Advisory Work per Analytics & Learnings

1995 — 2002 — 2009 — 2016 — 2021

Benchmarking analytics across over thousands of sites globally → learning for participating companies

TOTAL (Chemicals Only)

Since 2009, original participants in Analytics benchmarking model development

Predictive Analytics Hypothesis:

X = Excellence in Safeguard Improvement Capability

→ *Y = Excellence in Safety & OE Outcomes*

Safeguard Improvement Capability Analytics

ALL Business Processes that Enable Discovery and Repair of Weak Safeguards

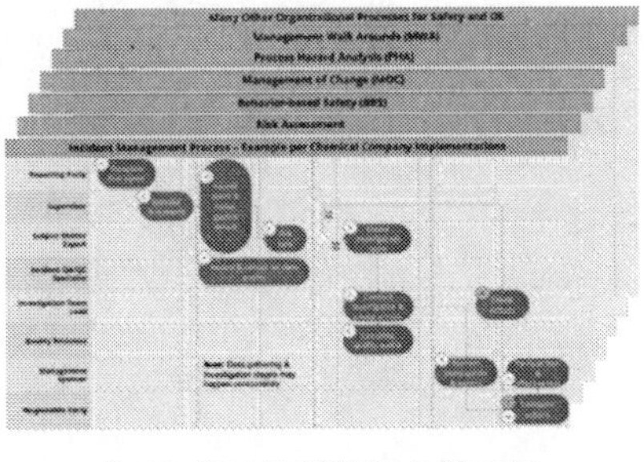

Process Execution Metrics to Measure Safeguard Improvement Capability

→

Maturity, Prevalence of Enterprise Software (EHS&S)

INTELEX
enablon
sphera
cority
BENCHMARK ESG | Gensuite

→

Industry Data Set and Predictive Analytics For Learning & Benchmarking

Industry Opportunity for Learning & Benchmarking

Global Industry Data Set & Analytics

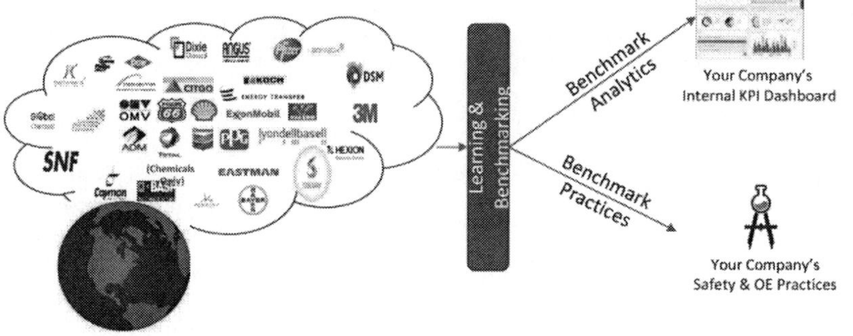

- World-wide prevalence of EHS&S Enterprise Software opens the door for huge collective analytics data set
- Benchmarking with peers - predictive metrics and lagging safety, process safety rates at operating site level
- Safety has unique industry sharing dynamic – not competitive – learn from top performers' practices to improve
- ESG – Environmental and Social Governance – is ultimate target of this benchmarking model – Safety & OE as first step

October 14, 2021 www.linkedin.com/in/joe-stough +1.832.526.9478 Slide 19

The journey for sustainable and innovative implementation of process safety big data analytics projects

Q&A

Co-organized by the Center for Chemical Process Safety and the European Process Safety Centre

VIRTUAL

2021 EUROPEAN CONFERENCE ON PROCESS SAFETY & BIG DATA

PERTAMINA HULU ROKAN CCPS EPSC

OCTOBER 13-14, 2021

Turning Cameras into Eyes – Success Stories and Lessons Learned From Leveraging Computer Vision to Improve Safety and Reliability in Facility Operations

Ir. Muhammad Riandhy Anindika Yudhy, S.T., M.Eng., I.P.M.
Field Engineer – Process Safety Management, PT. Pertamina Hulu Rokan

2021 EUROPEAN CONFERENCE ON PROCESS SAFETY & BIG DATA

VIRTUAL

OCTOBER 13-14, 2021 CCPS EPSC

Disclaimer: The materials provided in this presentation and any comments or information provided by the presenter are for information purposes only and nothing conveyed or provided should be considered legal

Co-organized by CCPS and ECPS

Why do we need Computer Vision in our operation?

PERTAMINA HULU ROKAN

An overview of Indonesia Upstream Sector and its operational challenges

- ~30,000 wells
- ~750,000 sq.km operating areas
- ~20,300 km pipelines
- ~30,000 km power lines & ~100,000 power poles
- ~100 gathering stations
- ~1000++ CCTVs in critical operations

https://www.skkmigas.go.id/infografis/fakta-umum-industri-hulu-migas

Such massive operations has its inherent operational hazards and threats.
Some of those hazards and threats posed major consequences to the operations, such as:

- Theft
- Illegal tapping
- Natural disruptions
- Public exposure to oil spill or leakage
- Weathering that leads to corrosion in equipment

Introduction to Computer Vision and How It is a Powerful Solution

- Computer vision (CV) is a field of artificial intelligence (AI) that **enables computers and systems to derive meaningful information** from digital images, videos and other visual inputs — and **take actions or make recommendations based on that information**.
- *AI enables computers to think, computer vision enables them to see, observe and understand*
- Surface facilities have multiple visual input devices ready to be exploited:
  - CCTV in gathering stations and offices
  - Cellphones
  - Drones
- Integration of CV into these devices will turn those visual devices into additional eyes that keep us operating safely and reliably despite the major threats
- 3 successful application of computer vision in surface facility operations that has been proven to improve operational safety and reliability despite the major threats inherently embedded into the surface facility operations.
- All applications are developed in-house by the employees, less involvement from third party

Case 1: Smart CCTV to detect possible theft in restricted area

PERTAMINA HULU ROKAN

Support Vector Machine Illustration

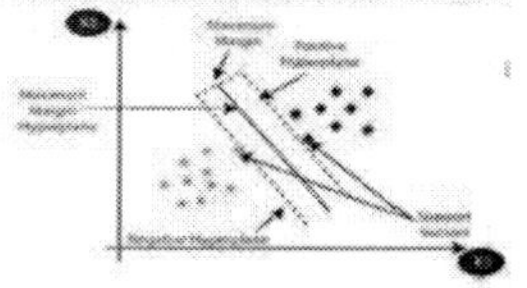

Histogram of Oriented Gradient Illustration

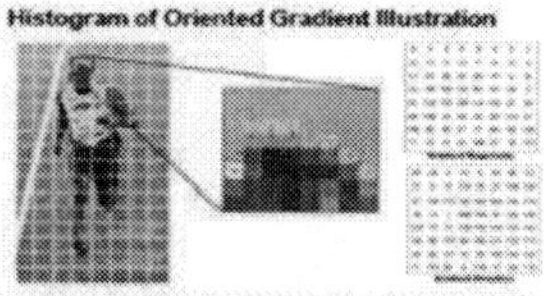

Create gradient from pixels of a picture and make a histogram of the gradient to extract features of the image for classification

Business Case

- 1000+ underutilized CCTV, only used for continuous manual monitoring by human. However, human has tendency to overlook critical things
- Opportunity to employ AI to analyze CCTV stream to detect people in restricted/limited Area and sending alert to security on duty.
- Improve security in critical facilities

Key Activities

- Develop AI based technique to detect people in selected CCTV stream.
- 2 approaches :
  - HOG (Histogram of Oriented Gradient) feature descriptor with SVM (Support Vector Machine)
  - Deep Learning Pre-Trained Model
- Sending the findings (snapshot and details) via messaging system (i.e., Microsoft Teams/Telegram)

CNNs for Classification: Feature Learning

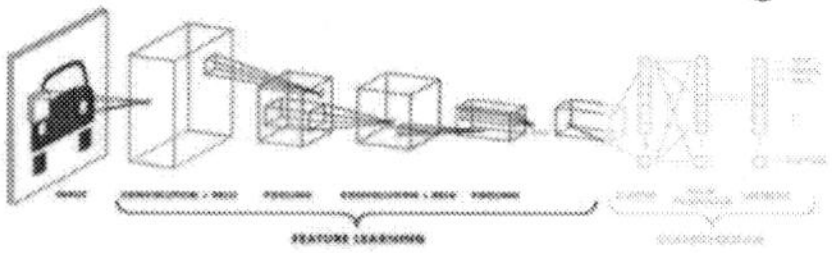

1. Learn features in input image through **convolution**
2. Introduce **non-linearity** through activation function (real-world data is non-linear)
3. Reduce dimensionality and preserve spatial invariance with **pooling**

CNNs for Classification: Class Probabilities

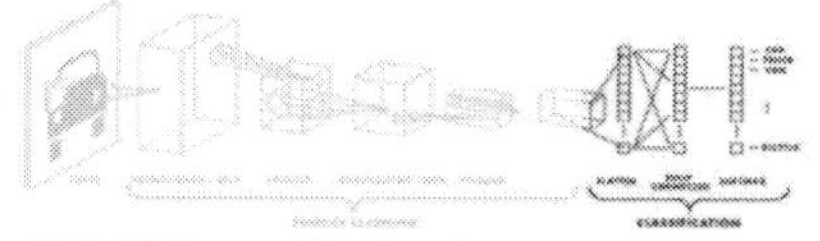

- CONV and POOL layers output high-level features of input
- Fully connected layer uses these features for classifying input image
- Express output as **probability** of image belonging to a particular class

$$\text{softmax}(y_i) = \frac{e^{y_i}}{\sum_j e^{y_j}}$$

Case 1 Process Improvement

PERTAMINA HULU ROKAN

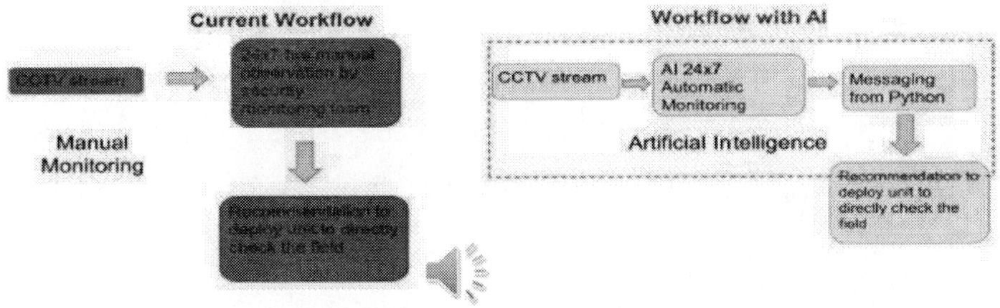

Process Overview

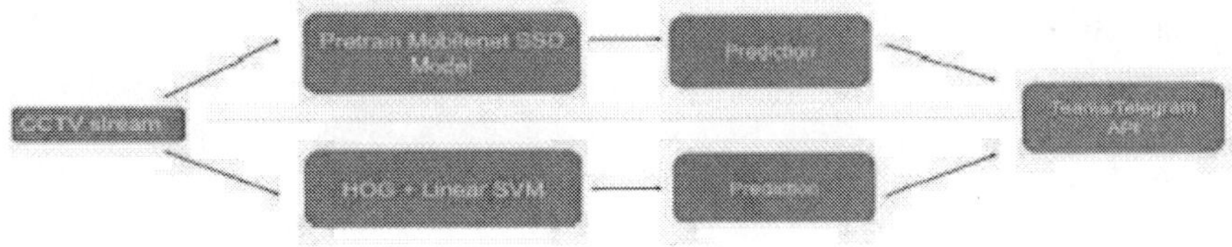

Case 1 Result

PERTAMINA HULU ROKAN

- Capability to detect people and sending alert to messaging apps.

Credits to Project Team:
- Ivan Susanto
- Ari Sukma Negara
- Ade Anggi Naluriawan Santoso
- Andika Putra

Case 2: Beyond Visual Line of Sight (BVLoS) drone with AI for pipeline surveillance

PERTAMINA
HULU ROKAN

Business Case

- Long shipping line that need to be monitored from gathering stations to end terminal
- There are illegal tapping cases found on shipping line that has impact to financial, security, and environment.
- Opportunity to utilize available drone for additional surveillance and monitoring to the current manual surveillance program.
- Object detection artificial intelligence algorithm is to be embedded into drone to detect suspicious activities in real-time.

Key Activities

- Drone technology assessment for pipeline surveillance
- AI model development to identify spatial changes to trigger alerts for any pipeline anomalies
- Permitting / exception process for beyond visual line of sight (BVLOS) and night flights mode from Government
- Proof of concept execution and review to support full scale implementation decisions

Case 2 Process Improvement

PERTAMINA
HULU ROKAN

BVLOS Drone + AI for Surveillance Architecture

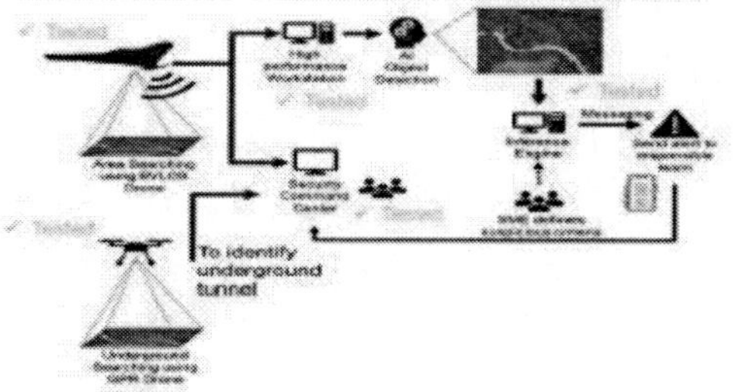

AI for Object Detection Process Flow

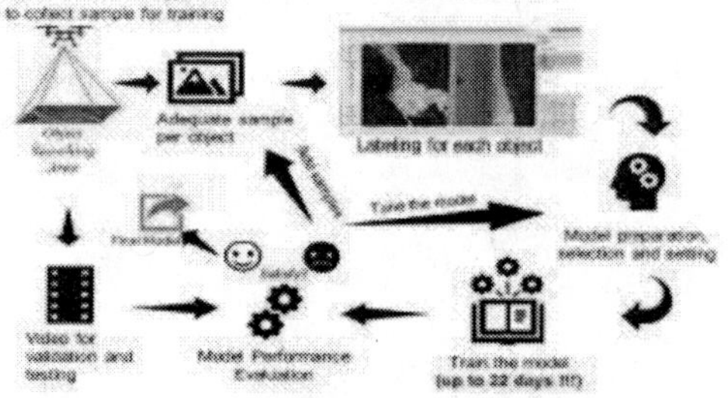

- Object detection
  - To detect 5 specific objects with unusual activity near pipeline: cars, motorcycles, people, cargo trucks and oil trucks
  - Use YOLO (CNN) and SSD Mobilenet
- Object segmentation
  - To classify areas for several objects : houses, roads and pipelines.
  - Use Auto-Encoder
- The two detection models are combined and then analyzed automatically based on human expertise (domain knowledge) to detect suspicious activity

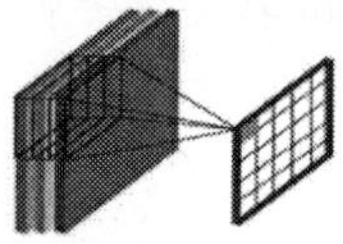

Convolutional Neural Network (CNN)
Use RGB kernel/filter to perform a convolution operation (moving matrix multiplication) with portions of input image to extract high level features of the image to be fed into neural network for classification

Case 2 Result

PERTAMINA HULU ROKAN

- ✓ No illegal tapping occurred during the Proof of Concept execution.
  - ✓ Drone surveillance was identified as one of contributing factor
- ✓ An AI Model with an automatic alert system has been built and tested.
  - ✓ The AI model was able to detect tanker trucks and people near the pipelines as suspicious activity and send alert via text messages (SMS)

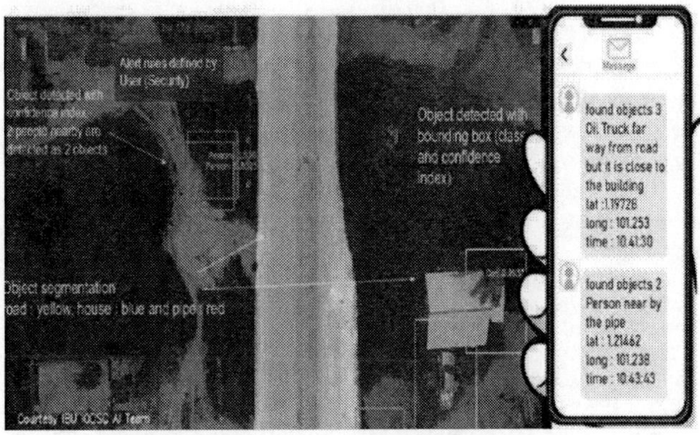

Credits to Project Team:

- Ivan Susanto
- Yusuf Hermawan
- Joko Nugroho Prasetyo
- Ramdhan Ari Wibawa
- Toto Arianto
- Deasy Paean
- Murdiyono
- Ahmad Sadikin
- M Attok Nur Isa
- Edi Supriyatno
- Fachrul Kamal
- Ray Ozma

Case 3: Drone with AI for powerline surveillance

PERTAMINA HULU ROKAN

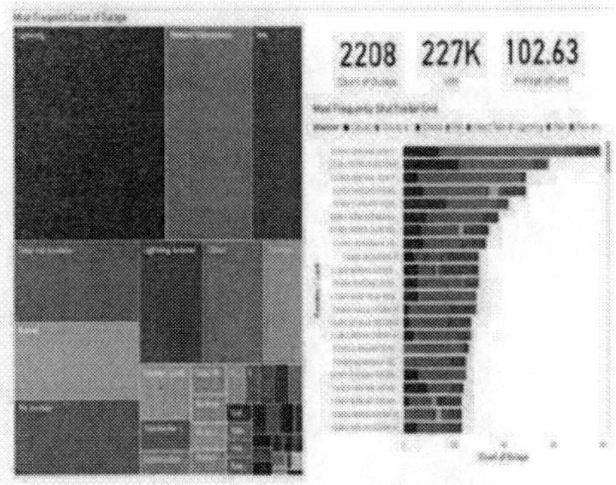

Business Case

- **Tree,** Animal**, Dead End Insulator** and **Pin Insulator** are among top 6 causes of power outage based on 5 years of historical data (2016 – 2021)
- Early detection of vegetation risk and insulator failure in critical power line to reduce the number of outage associated with vegetation, animal and insulator failure and improve power system reliability

Key Activities

- Drone technology assessment for power line inspection
- AI model development to identify insulator failure in power line and list out the location of the identified failure
- 3D modelling of power line and its surrounding to identify power line segment with vegetation risk
- Proof of concept (PoC) execution and review to support full scale implementation decisions

Case 3 Process Improvement

PERTAMINA HULU ROKAN

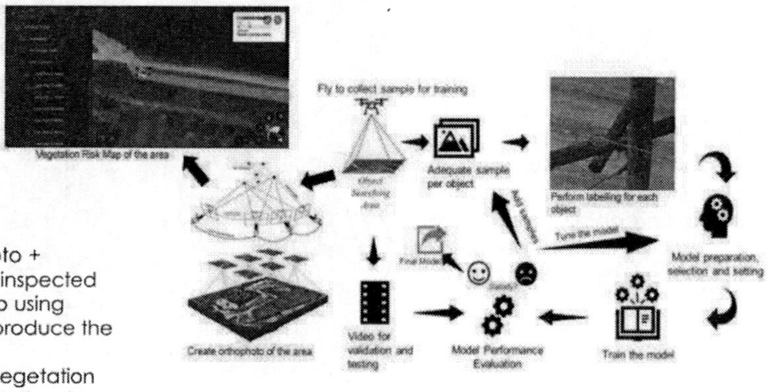

Approaches to powerline vegetation mapping

- ✓ Collect orthophoto (photo + coordinate) of the area inspected
- ✓ Combine the orthophoto using proprietary software to produce the 3D model of the area
- ✓ Insert a custom rule for vegetation
- ✓ Display location of vegetation risk relative to the nearest power pole for easier verification by Operations

Approaches AI model development

- ✓ Use YOLOv4 to develop the AI object detection model
- ✓ Detect 4 specific objects : bad insulator, good insulator, bad lightning arrester, good lightning arrester

Case 3 Result

PERTAMINA HULU ROKAN

- ✓ The AI model was able to differentiate good and bad condition insulator
- ✓ Vegetation risk map has been created for a pilot location where the list of vegetation risk identified by the model was verified by Operations.
  - ✓ 90% of the identified vegetation risk was confirmed as high-risk vegetation.
  - ✓ Operations was able to mitigate the risk early by taking action to trim the vegetation to safe height
  - ✓ 3 months after put on production, successfully reduce the number of outage and loss production opportunity due to vegetation to 0 in the pilot location.

Note: Good insulator ; bad insulator

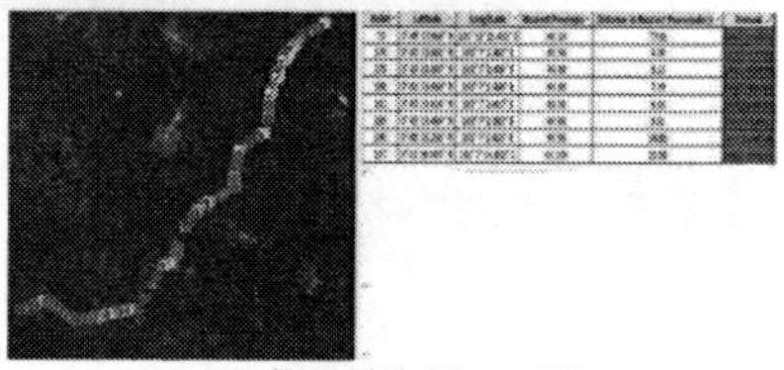

Note: High risk vegetation

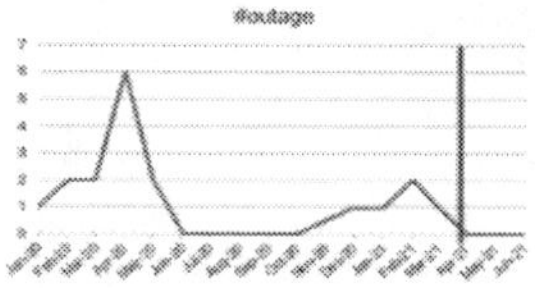

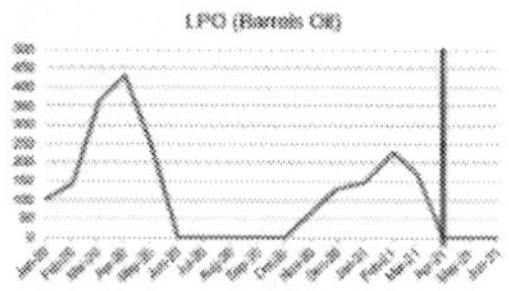

Credits to Project team

- Ivan Susanto
- Priambudi Pujihatma
- Joko Nugroho Prasetyo
- Ramdhan Ari Wibawa
- Soski
- Hendra
- Holmes Hotasiman Simbolon
- Andi Wirawan
- Ardiyansyah Lubis
- Muhammad Rismananda Pratama
- Muhammad Riandhy Anindika Yudhy
- Ginanjar Ihsan Prabowo
- Ray Ozma

Potential Expansion of CV Application

PERTAMINA HULU ROKAN

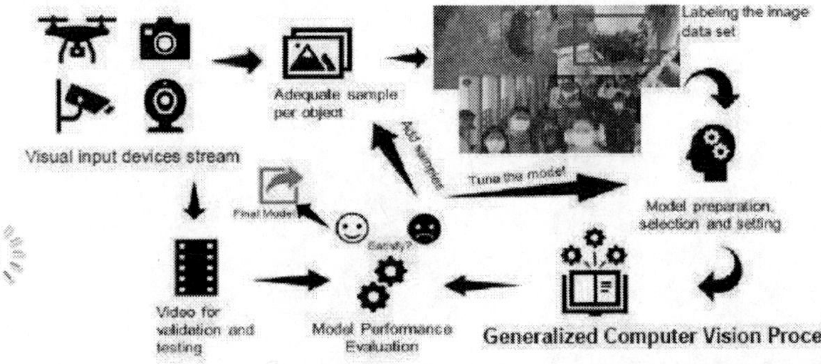

Generalized Computer Vision Process Flow

Potential Application using CCTV
- Physical distancing detection
- Virtual fencing
- Unsafe behavior detection (ii.e., mask & PPE usage)
- Vehicle Plate Number Detection
- Vehicle Speed Detection

Potential Application using drone
- Oil spill detection
- Corrosion detection in pipeline/powerline
- Inspection in hard-to-reach location (i.e., swampy area, tall structure)

Potential Application using Camera
- Product defect detection during Remote FAT
- Important variable detection and extraction from digital/scanned PSI document
- Using thermal cam, identify hot spot in powerline conductor

Summary

PERTAMINA HULU ROKAN

Challenges
- o Availability, quality and accessibility of training data for specific application (i.e., insulator failure data)
- o PC specification to perform computer vision work needs high specification processing power hence needs a full upgrade from the standard PC

Best Practices
- ❖ Combine domain expertise from functional team with AI/DS capability of IT to deliver innovative solutions for challenging operational issues (i.e., inspection in difficult location, theft detection, oil spill detection vegetation mapping, etc.)

Lessons Learned
- ✓ Application of computer vision algorithm to existing visual input devices has been proven to dramatically improved the effectiveness and efficiency of inspection and surveillance program
- ✓ Developing in-house computer vision application is relatively easy and it cost lower than using off the shelves market solution.

UNTUK INDONESIA
PERTAMINA
HULU ROKAN
TERIMA
KASIH
M.Riandhy
www.pertamina.com
135

Minimizing Human Error from Data Gathering in the Field

Paul Linkin - GOARC

13 Oct, 2021

GOARC

Introduction

- Process safety risk; complex and dynamic
- Increasing quantity and quality of data from measurement devices to Control Rooms
- Supplementary data gathered in by personnel is still vital
- Humans make errors
- Understand and manage error likely situations; error tolerance
- Technology can enable/support data collection integrity
- Potential challenges

2 | ©2021 GOARC

GOARC

Humans Errors in Data Gathering

- Misreading
- Misinterpretation
- Miswriting/typing
- Memory lapses
- Incorrect sequence
- Partial completion
- Fabrication of results
- Etc.

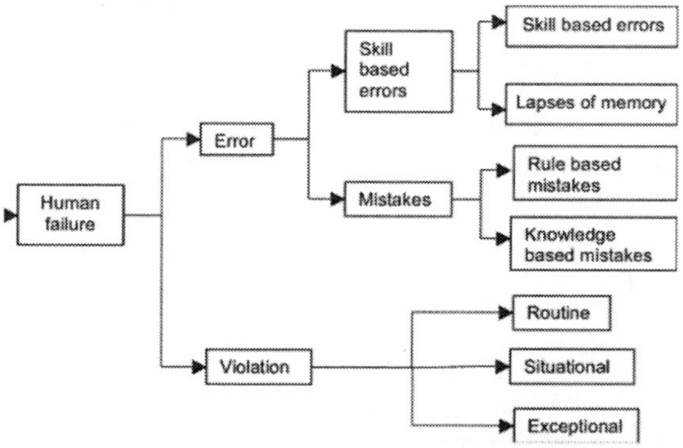

GOARC

Factors in Data Gathering Error (Context)

- Task/Environment Factors
  - Environmental Conditions (TEF)
  - Procedures and Guidelines(TEF)
  - Ergonomics (TEF)
  - Task Time and Duration (TEF)
  - Supervision (TEF)
  - Etc

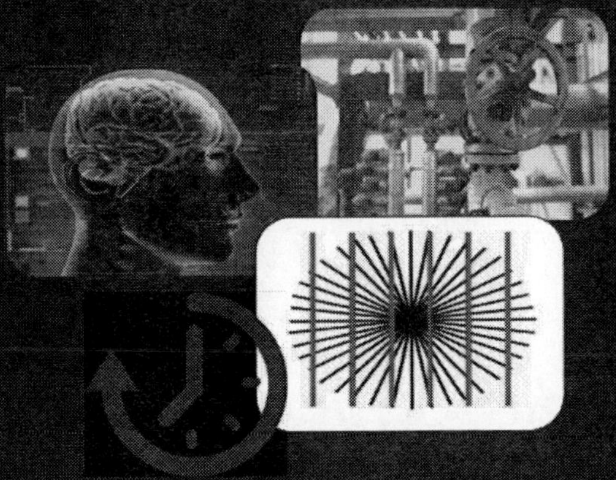

- Human Factors
  - Physical capability (HF)
  - Skill/Knowledge (HF)
  - Fatigue (HF)
  - Perception (HF)
  - Etc

GOARC

Technology in Data Gathering

GOARC

Technology: Portable/WearableDevices

- Connected Worker; bi-directional (who)
- Tech solutions in one device;
  - GPS location (where)
  - Camera/Video(what/how)
  - Guidance/Procedures
  - Automatic readings
  - Real Time Communication
- Productivity Benefits
  - Tool Time Efficiency
  - Right first time

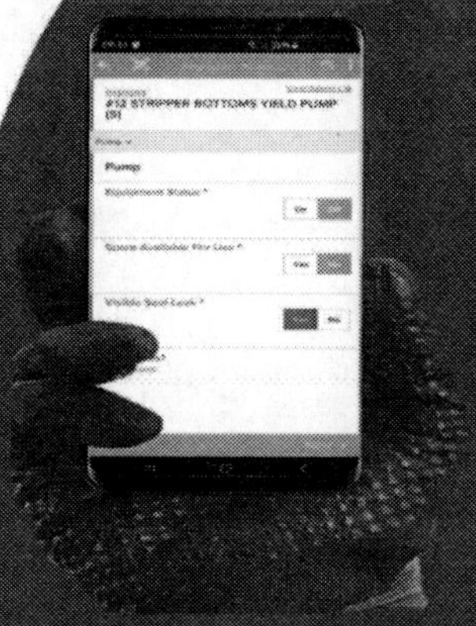

Technology: Software

- Apps/Cloud-based platforms
- Dynamic/conditional routines
- Data gathering format options
- Command center consolidation
- Data visualization on dashboards
- Open-source platform interfaces with Systems/IOT Devices
- Interpretation and processing (AI/ML)

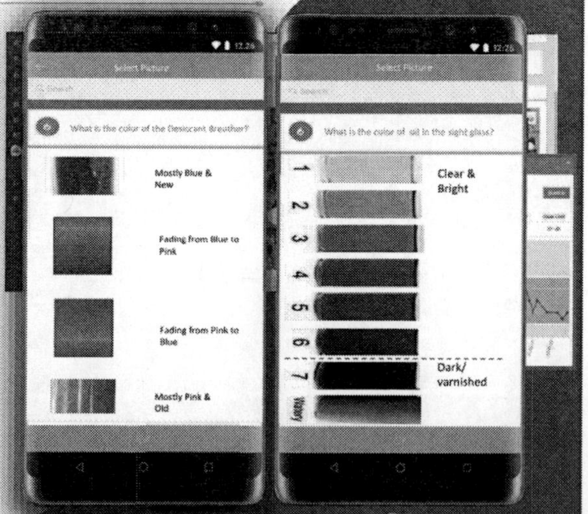

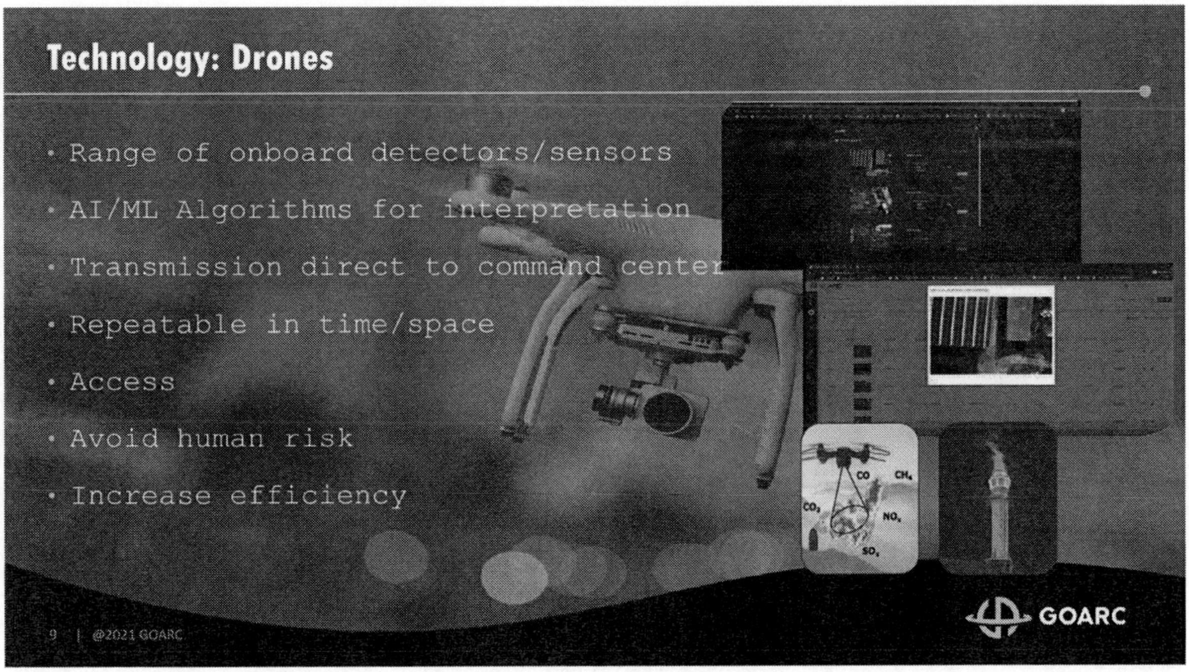
Technology: Drones
• Range of onboard detectors/sensors
• AI/ML Algorithms for interpretation
• Transmission direct to command center
• Repeatable in time/space
• Access
• Avoid human risk
• Increase efficiency
9 | @2021 GOARC
GOARC

Technology: Connectivity/Networks
• WiFi, Private Networks
• Coverage
• Bandwidth
• Interference
• License requirements
10 | @2021 GOARC
GOARC

Challenges with Technology

- Availability
- Reliability
- Consolidation
- Utilization

GOARC

Benefits and Value

- Broad Risk-Based PSMS benefits
- Improved data gathering integrity
- Increased capacity for dynamic process risk assessment
- Data availability - right people; right place; right time
- Enhanced risk-based decision making
- Operator Safety
- Productivity gains

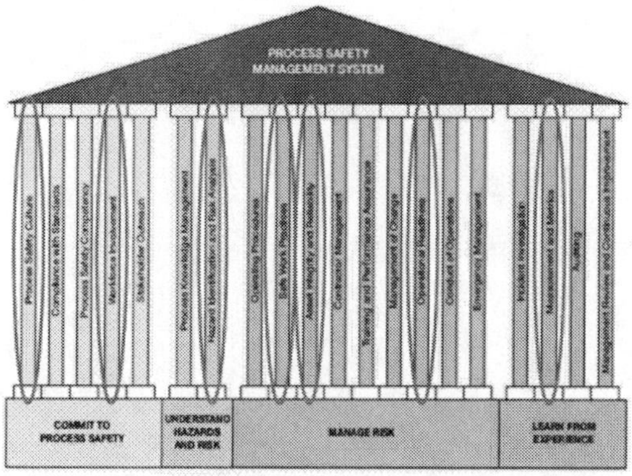

GOARC

Summary

- Personnel still vital to data gathering for process safety
- Humans make errors; influenced by contributing factors
- Need to understand the context of error and data gathering needs (organizational and human factors)
- Thoughtful implementation of technology(hardware/software) can enable reduction of many error-likely situations and error types
- Consider and address potential limitations and challenges
- Benefits in data integrity, worker safety, asset integrity, plant availability and productivity

GOARC

Thank you

- **Paul Linkin**
  - paul.linkin@go-arc.com
  - polinki13@gmail.com
  - +1 832 499 6033
  - Houston, TX USA - CST (UTC-6)
- **Jim Petrusich**
  - jim.petrusich@go-arc.com
  - +1 503 913 1337
  - Portland, OR USA - PST (UTC-8)

GOARC

VIRTUAL

2021 EUROPEAN CONFERENCE ON PROCESS SAFETY & BIG DATA

CCPS EPSC

OCTOBER 13-14, 2021

2021 EUROPEAN CONFERENCE ON PROCESS SAFETY & BIG DATA

VIRTUAL

OCTOBER 13-14, 2021

CCPS EPSC

Neural Survival Modeling to Predict Process/Environmental Excursions in an Operating Facility

KJ Moon, PhD
Adv. Data Scientist
Honeywell Connected Enterprise

Tony Downes
Global Process Safety Advisor
Honeywell

Co-organized by CCPS and ECPS

Background/Motivation

- The environmental emission – as well as the rest of the operational processes – is monitored with threshold-based alarming
- This alarming mechanism inherently doesn't provide much lead time for operators to find causes and perform corrective actions
- There is a rich set of time-series data collected for years

Question:

Could we leverage this data and develop a data-driven model to provide an early-warning of elevated probability of an environmental emission with significant lead time?

Data

- 2-year process historian data from an operating plant, with 6-minute rate
- For the study, 39 PHD params or "tags" were used
- Environmental surge is monitored with two params – Component A and Component B rolling averages
- Older 70% of the data used for training model, and the other 30% for testing

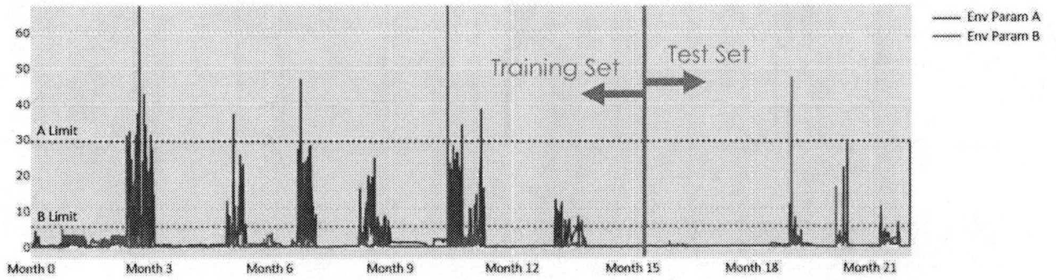

Modeling Approach

- Question to answer: "from now (T), how likely would the alarm occur within the next X of time?"
- Survival analysis* model was developed to answer this kind of question
- It can answer to other similar questions like
  - Probability that the event may <u>not</u> occur for period X from now
  - Assuming the event didn't occur for T+X, probability to occur at T+X+1 (next time period)
- The original survival model framework is with linear modeling but can also be extended to a nonlinear form. The modeling approach will look like below:

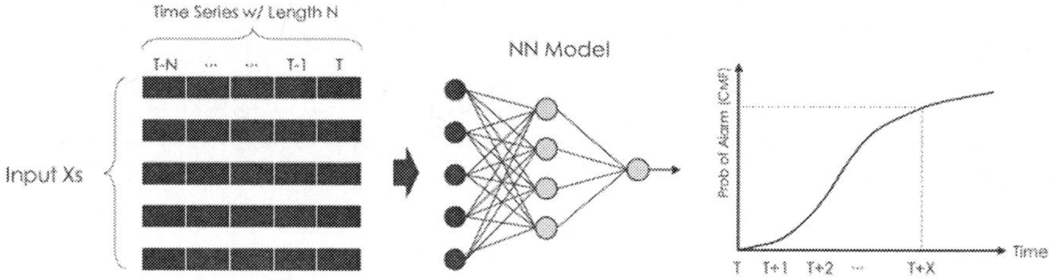

* Survival analysis is a branch of statistics for analyzing the expected duration of time until one event occurs, such as death in biological organisms and failure in mechanical systems.

Modeling Workflow

Raw Data
- 2-yr span
- 6-min data rate
- 39 tags
- Alarm locations found with thresholds

Feature Transformation
- Aggregate to daily max
- Generate "X & Y"
- X's TS length = 5 days
- Split into training & test set

Model Training & Testing
- Model Selection
  - Linear
  - RF
  - Feed-Fwd NN
  - CNN
- FF-net as final choice

Trained Model

Output:
Prob of alarm occurring within next 5 days

Result

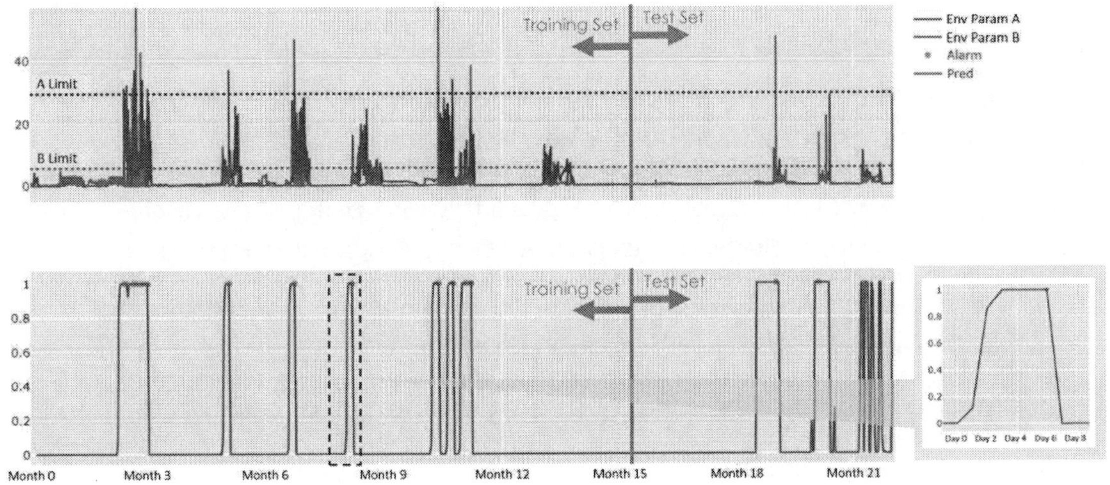

Result: Close-Up on Test Set

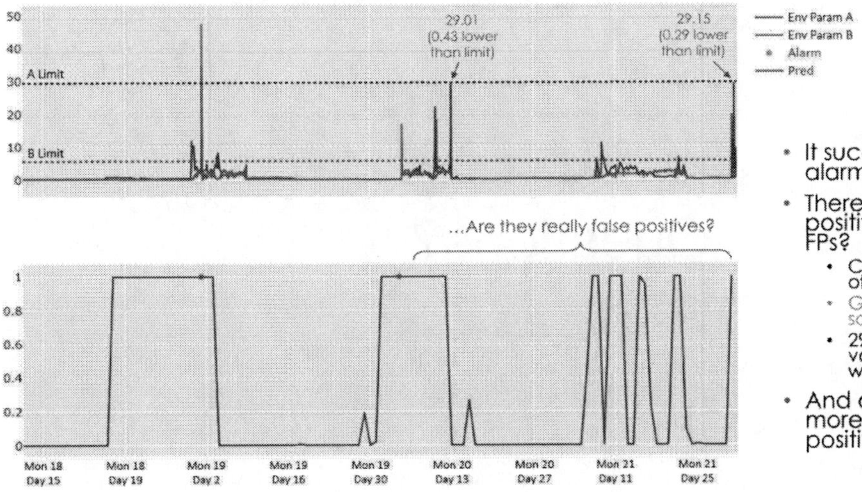

- It successfully predicted the alarms
- There are also false positives, but are they really FPs?
  - Challenge is at the ambiguity of the alarm
  - Good idea to flag to the plant so they can intercede
  - 29.01 is lower than the limit value 29.44, but do you really want pass it??
- And as always, we need more data (especially, the positive cases)

Summary

- Applied survival modeling to predict the probability of alarm occurring in days ahead
  - Operator may even be able to prevent the alarm from occurring
- The result seems quite promising, even with some false positives
- Obviously, the model is not production-ready. We need more rigorous validation, test, and model tuning. To do it, more data is needed
- It could be a good piece of the puzzle for building advanced alarm management. Other pieces are:
  - Earning trust: good model accuracy is just a basic condition
  - Explainability: causal identification is on the work
  - Culture & system changes: operators act on training, manual or protocol. Model adaption needs proper change of and integration with the rest of the system

2021 EUROPEAN CONFERENCE ON PROCESS SAFETY & BIG DATA

OCTOBER 13-14, 2021

CCPS EPSC

VIRTUAL

Questions?

Co-organized by CCPS and ECPS

empirisys

DETECT

A Tool to Improve Loss Prevention from Human Error in High Hazard Industries through the Identification of Error Traps Using Machine Learning

2021 EUROPEAN CONFERENCE ON PROCESS SAFETY & BIG DATA

CCPS EPSC

About Empirisys

empirisys

We use engineering & operational expertise and machine learning to create actionable insight to help complex, high hazard organisations manage their risks

About the Company

Established in October 2020 by two co-founders

- Gus Carroll - former Chief Engineer at Centrica (£30bn end to end energy company)
- Pete Sueref - former Data Science Director at Centrica
- We have a team of highly experienced operational & engineering professionals and PhD level data scientists

What We Do

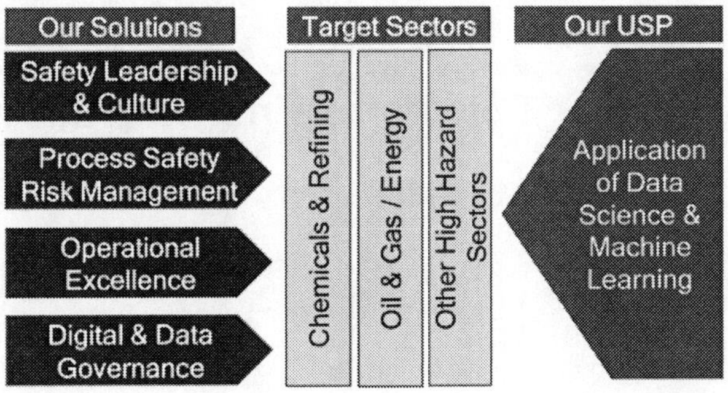

2

empirisys

Gas from the wellbore erupts through the rig floor which ignites an explosion killing 11 and injuring 17

Transocean & BP Oil Rig
Gulf of Mexico
April 20, 2010

3

empirisys

DETECT

Detection of Error Trap Effect and Consequence Tool

4

What is DETECT and What Does It Do? empirisys

- DETECT helps to identify the organisational weaknesses that lay "hidden in plain sight" at every operational asset or facility
- These are often called "Error Traps" or "Performance Influencing Factors"

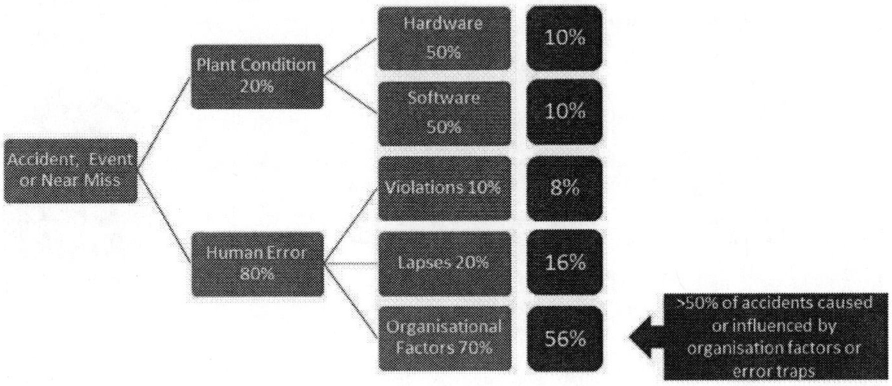

5

What is an Error Trap? empirisys

- Result of organisational factors and latent defects that increase the likelihood of human error
- In safety critical settings, this can result in catastrophic consequences
- Error traps are often "hidden in plain sight" but are extremely difficult to spot

JOB	• Inappropriate procedures • Complex task instructions • Working environment
PERSON	• Stress & morale • Fatigue • Physical capability & condition • ...
ORGANISATI ON	• Work pressures (production vs safety) • Psychological safety • Communication • ...

HSE Health & Safety Executive

6

DETECT v0.01 Prototype

empirisys

DETECT brings a new functionality to the domain of safety by leveraging machine learning to make sense of vast amounts of organisational data, rather than relying on existing qualitative techniques.

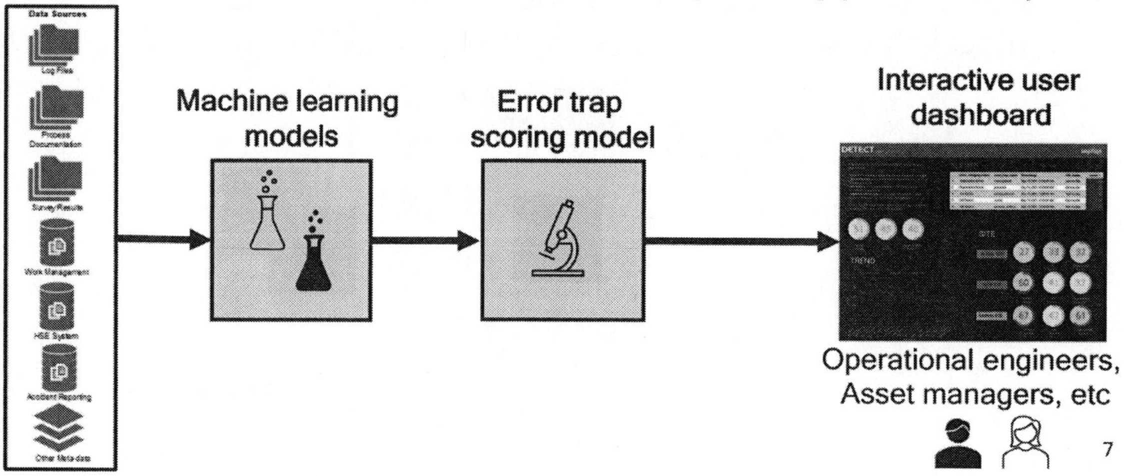

7

Main Challenges

empirisys

- Is it possible to extract, cleans and transfer unstructured data from operational technology sources such as maintenance systems, HSE reporting systems etc? **Data Feasibility**
- Is it possible to detect error traps automatically from data? **Machine Learning Feasibility**
- Do users need our product? **Desirability**
- Will user believe the output of the product? **Credibility**

8

Data Feasibility

empirisys

STRUCTURED DATA

REFLECTS...	EXAMPLES...	WHAT MACHINE LEARNING OFFERS
• Formal structure & process • The way we say things get done • How the world sees us	• DCS/PI • Alarms & alerts • Action closures • Risk ratings • Financial KPIs etc	• Correlate trends with future outcomes • Early warning • Feedback / action loops

UNSTRUCTURED DATA

REFLECTS...	EXAMPLES...	WHAT MACHINE LEARNING OFFERS
• Informal structure & process • The way things really get done • The stories we create • The unexplored way we are • The true culture of the organisation	• Worker logs • Observations • Interviews • Surveys • Meeting notes • Audit reports etc • Emails • Social networks • Conversations etc	• Themes & topics • Sentiment • Relationships • Networks • Clusters

- Unstructured data held in different formats on IS platforms that do not easily connect
- Often use different rulesets for key metadata such as timestamps, tags and location.

9

Machine Learning Feasibility
High Level Process

empirisys

Collate & Clean Source Data → Extract Data Using ML Models → Calculate Word Importance → Build Correlation Analysis → Visualise Data by Network Analysis

Collate & Clean Source Data
- Unstructured text data
- Incomplete syntax
- Sometimes having typos
- Different formats for common data sets (time, date etc)

Extract Data Using ML Models
- Convert to lower case
- Remove non-alphabets
- Remove common words
- Word stemming
- Apply gram model analysis (unigram & bigram)

Calculate Word Importance
- Use various statistical models
- Term frequency analysis
- Inverse document frequency analysis
- Weighted term importance

Build Correlation Analysis
- Pearson correlation coefficient
- Compute for all terms within entire corpus
- Determine strength of correlation

10

Machine Learning Feasibility - Results

empirisys

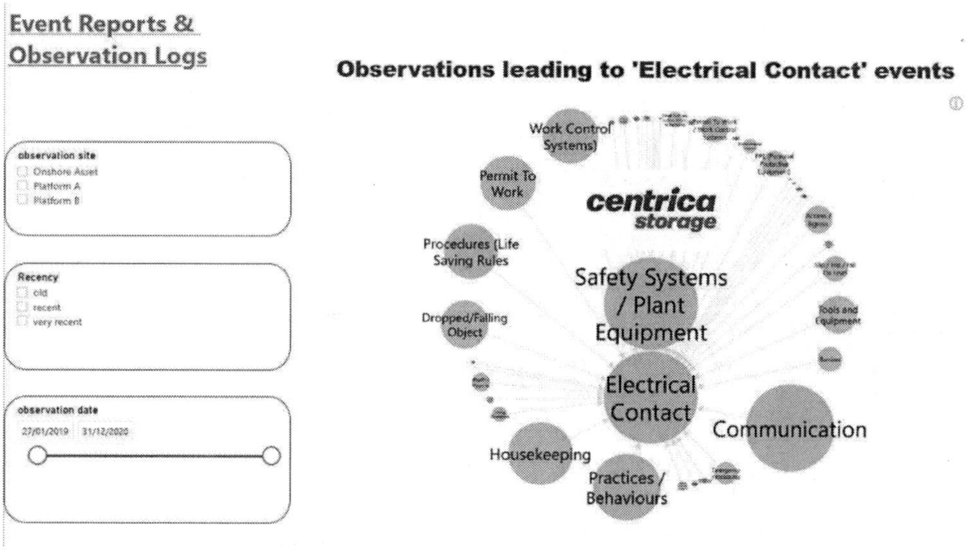

11

Desirability Testing

empirisys

- 5 subject-matter experts in high-hazard industries and process safety
- 1-hour interview with interactive prototype (Google Data Studio dashboard)
- User experience and desirability questions

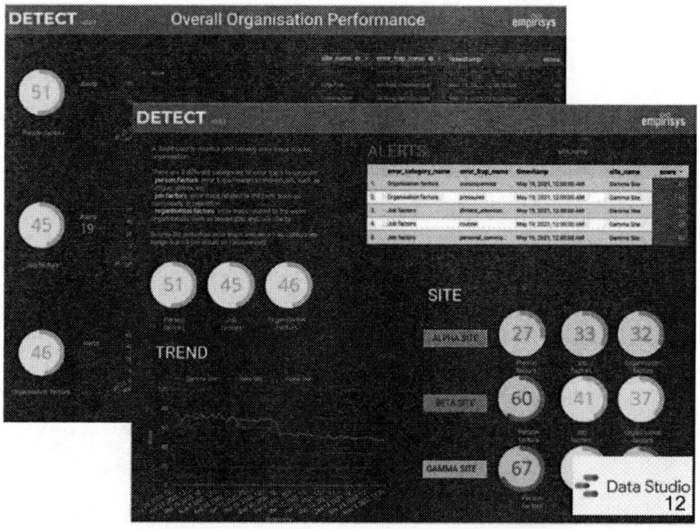

12

Usability Study - Results

empirisys

All 5 experts interviewed felt there was a need for a product like DETECT

Key Observations

- Tool requires a minimum level of data maturity from the organisation
- Most useful when embedded into existing risk management and process safety tools
- Data evidence for issues is invaluable

Key Recommendations

- Favour simplicity
- Be able to take action and lead to interventions
- Need for transparency

13

DETECT is Currently in Development

empirisys

Successful initial user desirability testing

Error trap-marker links establishe d

Machine learning feasibility tested

Operational data available

centrica storage

• • •

Can we do this at scale?

DETECT Prototype development - feasibility tested	DETECT MVP launched with partner companies - extensive testing phase	Production-ready version of DETECT launched	Expand to new customers within target sector	Expand to other high-hazard industry sectors
Q1 - Q3 2022	Q4 2022 - Q1 2023	2023 - 2024	2024-2025	2026+

14

empirisys

Thank You

Website: https://empirisys.io
Presenter contact (Nyala Noe): nyala@empirisys.io
General enquiries: info@empirisys.io

15

ioMosaic®
Minimizing risk. Maximizing potential.®

2021 EUROPEAN CONFERENCE ON PROCESS SAFETY & BIG DATA
CCPS EPSC

October 13-14, 2021 | Virtual European Conference on Process Safety & Big Data

Asset Integrity Training and Workflows to Avoid Human Errors

Dianne Coon, CSP
Principal Consultant
coon.d.nh@iomosaic.com

QMS_7.3_7.4.F12 Rev.4

ANAB ISO 9001:2015

Mechanical or Asset Integrity Process Safety Regulations

- Mechanical or asset integrity is a key part of process safety and is required worldwide
  - Seveso III, Annex III(b)(iii) Operations Control including Maintenance (Europe)
  - 29 CFR 1910.119(j) Mechanical Integrity (US)
  - AQ/T 3034-2010(4)(4.7) Mechanical Integrity (China)
  - Many other standards such as ISO, OSHA, etc.

ioMosaic® 2

Four common barriers to proper implementation of a successful asset integrity program

- Lack of management support
- Lack of training
- Lack of systems for managing inspections and deficiencies
- Lack of metrics and reports to monitor progress for continuous improvement

Management needs to provide structure and resources for implementation

- Develop corporate policies and procedures to include:
  - Scope
  - Roles and responsibilities
  - Process for equipment selection and risk prioritization
  - Identifying damage mechanisms and failure modes
  - Documents for appropriate inspection, testing, and preventive maintenance (ITPM)
- Provide adequate training and resources for implementation
  - Engineering
  - Maintenance
  - Contractors (if inspections are outsourced)
- Metrics to monitor implementation and support corrective actions needed for improvement

Potential consequences of inadequate training

Consequence from human error	Due to Lack of Training on…
Maintenance activities not executed	roles and responsibilities
Equipment failure	damage mechanisms and missing ITPM that was needed to detect those mechanisms
Critical equipment not placed on the equipment list which leads to failure	key information that should be utilized to select the critical equipment and risk prioritization criteria
ITPM not completed appropriately which leads to failure	maintenance procedures
New equipment fails at startup	quality assurance procedures for the proper fabrication, installation, and testing of new equipment

Systems needed to manage ITPM tasks and correct deficiencies

- Electronic systems should be implemented to manage ITPM tasks:
  - Documenting equipment process safety information
  - Scheduling ITPM on time
  - Executing ITPM per standards and procedures
  - Documenting and reviewing ITPM results
  - Determining equipment fitness for continued use
  - Updating ITPM tasks and frequencies
  - Reporting ITPM or deficiency resolutions that are due, late, or upcoming
- Electronic system should also address deficiency resolutions
  - Temporary and permanent repairs
  - Mitigation and Fitness for use evaluations
  - Removing equipment from service

Example Forms to Record Process Safety Information

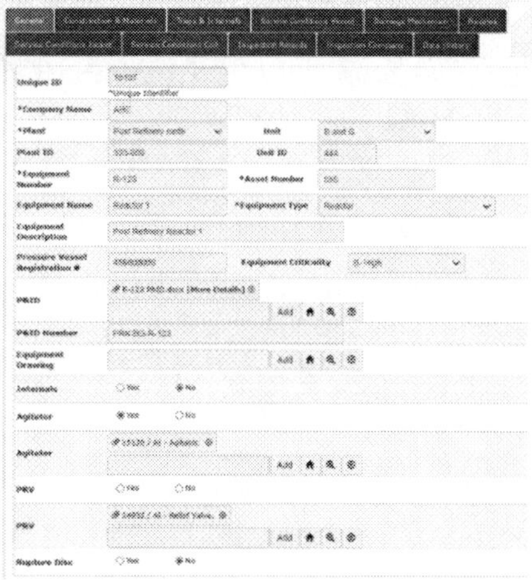

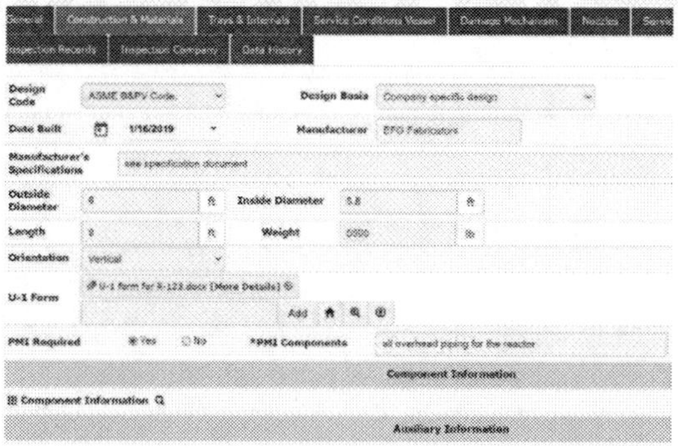

Source: Process Safety Enterprise® - ioMosaic Corporation

Example Asset Integrity Workflow

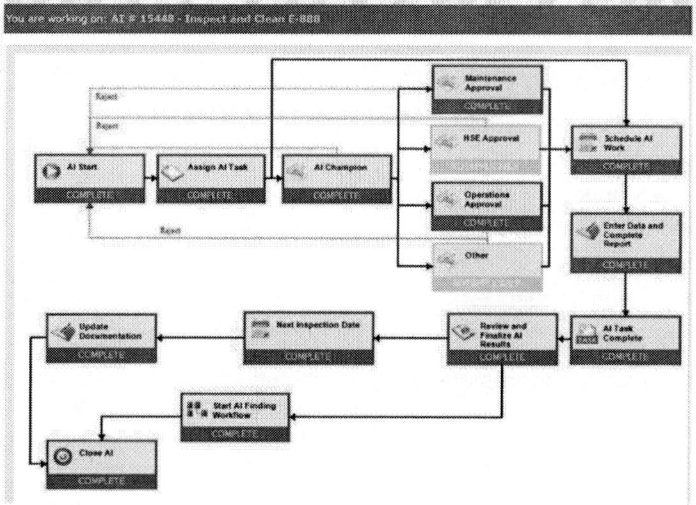

Source: Process Safety Enterprise® - ioMosaic Corporation

Example Asset Deficiency/Finding Workflow

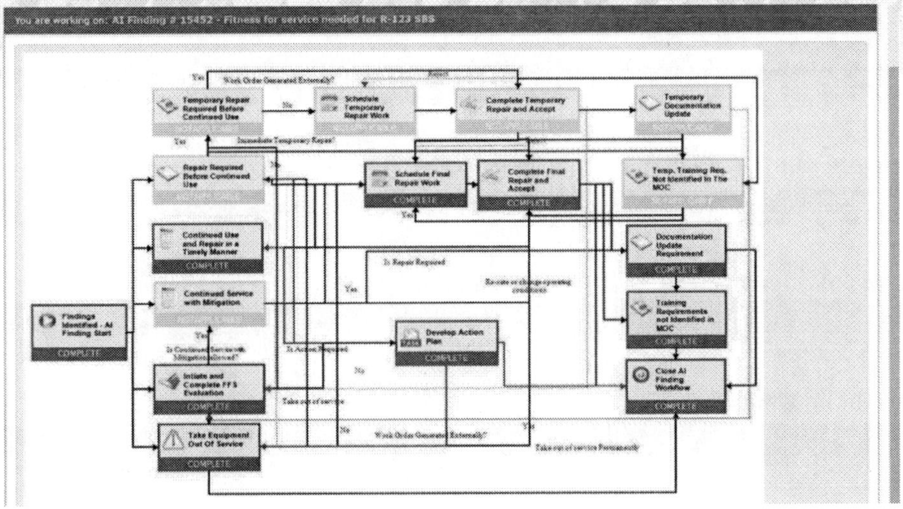

Source: Process Safety Enterprise® - ioMosaic Corporation

Potential consequences of inadequate workflow

Consequence from human error	Due to lack of workflow to establish…
ITPM is not appropriate	PSI not recorded and improper industry standard used to establish ITPM
Equipment fails due to extreme conditions	PSI for service condition limits
ITPM or resolution not executed	reports to show tasks that are due
Deficiencies not addressed appropriately which leads to catastrophic failure	appropriate resolution of deficiencies and a workflow to ensure timely and complete resolution
Equipment fails before next scheduled ITPM	review by qualified individual to assess equipment condition & update ITPM interval

Report examples

Inspections due report

Equipment Number	Form Name	Equipment Type	Inspection Type	Next Inspection Date ▼	Plant Site	Unit Name	Equipment Criticality	PMI?	Operation Status	Inspection workflow started?	AI Due Date
pump	13222 / AI - Pressure Vessel	Storage Tank	Formal Visual External	1/3/2022	Dispersed Size	B and G	A - Very High	No	Full production	No	N/A

Inspections overdue report

Equipment Number	Form Name	Equipment Type	Inspection Type	Next Inspection Date ▼	Plant Site	Unit Name	Equip Criticality	PMI?	Operation Status	Inspection workflow started?	AI Due Date
A-123	15120 / AI - Agitator	Agitator	Vibration	1/18/2020	Hard Resin Kettles	B and G	B- High	No	Full production	AI Workflow 15122	08/27/2020

Deficiencies/Findings due report

Equipment Number	Equipment Type	Description	Disposition	Ready Task	Current Responsible Person	Final Resolution Date ▼	Criticality	Findings Workflow Link	Plant Site	Unit Name
HX-123	Heat Exchanger - Shell and Tube	FFS evaluation required	Fit-For-Service Evaluation	Intiate and Complete FFS Evaluation	Dianne Coon	11/4/2021	B- High	18093 / Asset Integrity Finding	CTO/Hill track	CTO Plant

Deficiencies/Findings overdue report

Equipment Number	Equipment Type	Description	Disposition	Ready Task	Current Responsible Person ▼	Final Resolution Date	Criticality	Asset Integrity Finding Form Name	Plant Site	Unit Name
EV-888	Emergency Vent	test	Repair Required Before Continued Use	Repair Required Before Continued Use	Tao Yao	9/23/2020	B- High	15503 / Asset Integrity Finding	Dispersed Size	Lignin Production

Source: Process Safety Enterprise® - ioMosaic Corporation

Summary

- Management needs to provide support and resources to implement a successful asset integrity program
- Adequate training and workflows are needed to prevent human errors that can lead to catastrophic events
- Workflows are needed for
  - executing the equipment ITPM
  - addressing deficiencies
  - reporting on tasks that are due, late, or upcoming

Source: ioMosaic Stock Photography

About ioMosaic Corporation

Through innovation and dedication to continual improvement, ioMosaic has become a leading provider of integrated process safety and risk management solutions. ioMosaic has expertise in a wide variety of areas, including pressure relief systems design, process safety management, expert litigation support, laboratory services, training, and software development.

ioMosaic offers integrated process safety and risk management services to help you manage and reduce episodic risk. Because when safety, efficiency, and compliance are improved, you can sleep better at night. Our extensive expertise allows us the flexibility, resources, and capabilities to determine what you need to reduce and manage episodic risk, maintain compliance, and prevent injuries and catastrophic incidents.

Our mission is to help you protect your people, plant, stakeholder value, and our planet.

For more information on ioMosaic, please visit: **www.ioMosaic.com**

RISK ALIVE

SMART RISK DECISIONS DRIVEN BY DATA

USING BIG DATA TO GENERATE CERTIFIED LAB PHA

SAVING TIME AND INCREASING RISK DISCOVERY

PRESENTED BY ROXANNE MAI

PROCESS SAFETY ENGINEER IN TRAINING

2021 EUROPEAN CONFERENCE ON PROCESS SAFETY & BIG DATA

CCPS EPSC

SPRING22 +18TH GCPS

A Joint AIChE and CCPS Meeting

AGENDA

- PHA Overview
- Challenges
- Solution - Certified Lab PHA
- Benefits

PHA Overview

Purpose
An organized effort to identify and evaluate hazards associated with processes and operations

Team
A PHA team is formed consisting of competent members with relevant skills and experience

Scope
Select an appropriate scope which can be analyzed by the team to align with the design intent

Hazards
Using the collective knowledge in the room to identify hazards and risks

Report
Capture the cooperative thinking of the team in reports to preserve the outcomes of the session

Why Process Hazard Analysis (PHA)

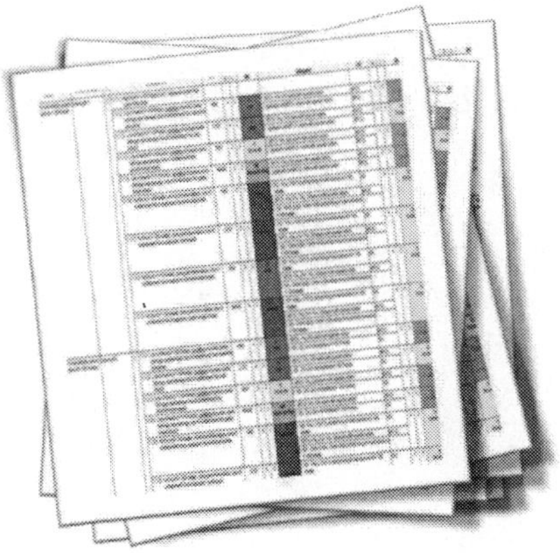

Scalability

Ease to Execution

Challenges

To increase PHA quality, completeness, and consistency, there are some items to consider

- Availability of Subject Matter Experts (SMEs)
- Labor intensive to create and maintain
- Inconsistent formats
- Simplifying ease of use for facilitators

Solutions

Certified Lab PHAs

feedstock for the PHA session to bring in knowledge from across the industry

- Large PHA data repository
- Powerful Machine Learning applied
- Risk Discovery on ensuring data quality

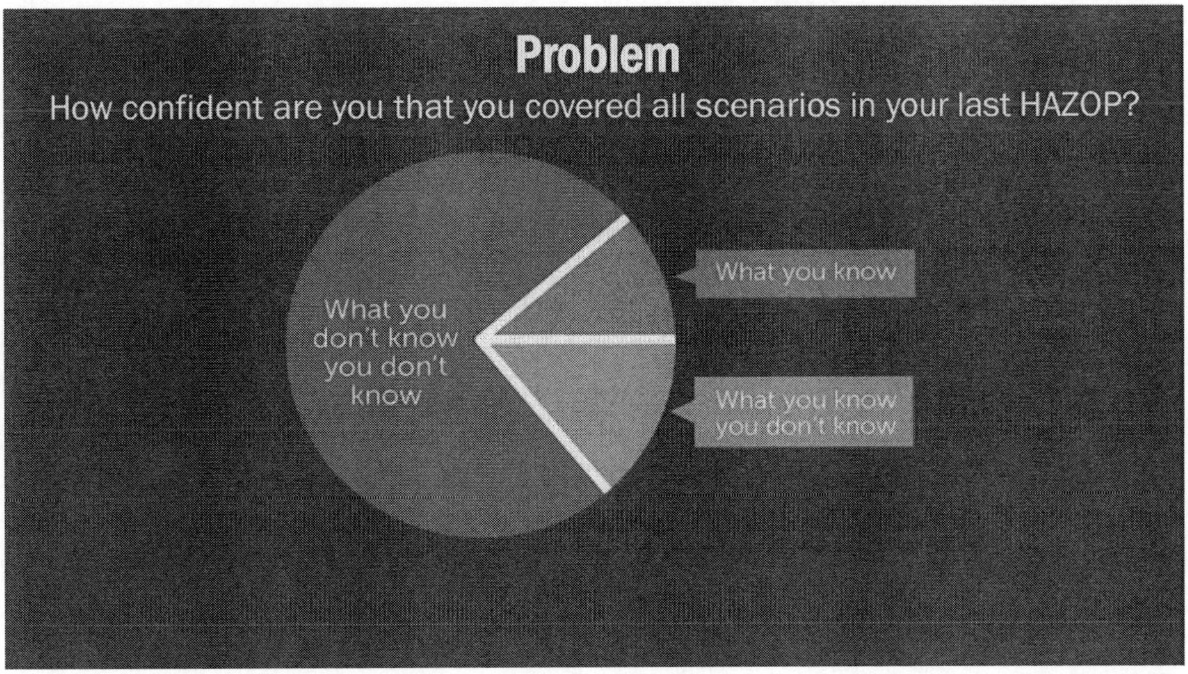
Problem
How confident are you that you covered all scenarios in your last HAZOP?
What you don't know you don't know
What you know
What you know you don't know

To validate how well a PHA covered all applicable risks

Risk Discovery

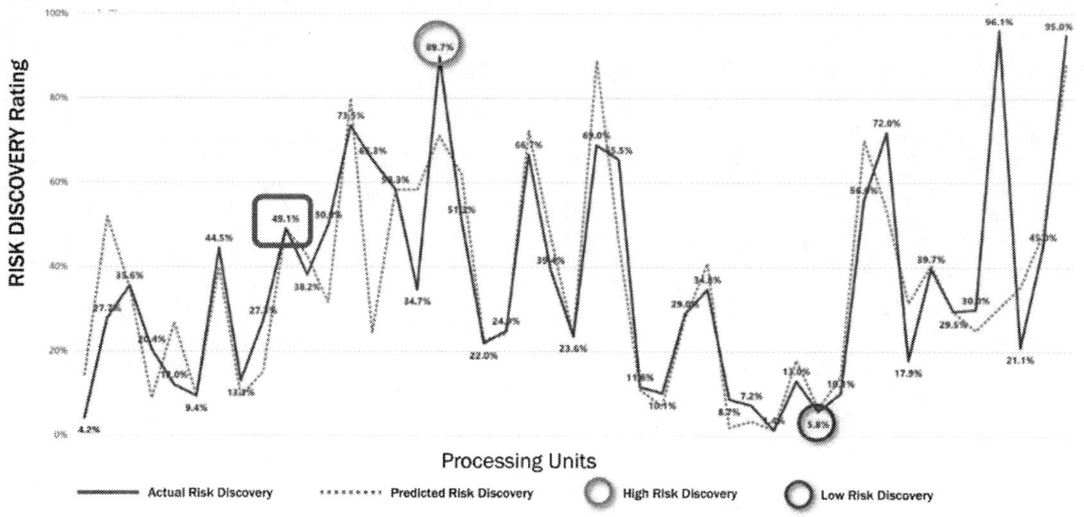

What is Certified Lab PHA?

A consultants and/or internal PSM exports' tool

Data driven using industry information

Leveraging the Industry-wide commonly analyzed hazardous scenarios by equipment

Make sure we don't miss anything

Where does the Industry data come from?

The Worlds Largest Process Hazardous Scenario Library

Processing Units	Pieces of Equipment	Unit Types	Contributing Companies	Countries
4,760	**262,000**	**300**	**161**	**12**

Learn from Industry

Metrics as of August 20, 2021

Consultant Tool – Risk Alive Certified Lab PHAs

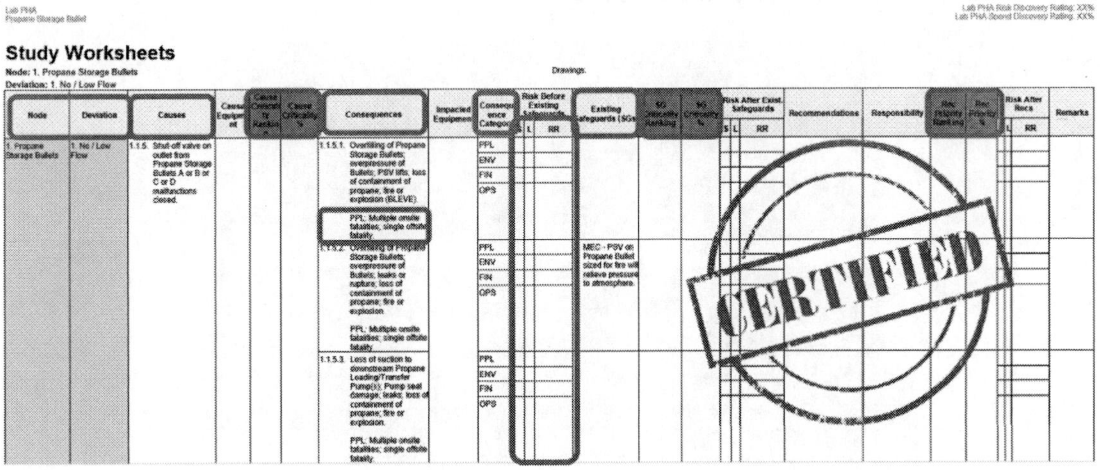

Data-driven Certified Lab PHAs – Powered by Risk Alive

Industry-wide commonly analyzed scenarios into clients PHA Template

Benefits – Time Saving

REDUCE PREPARATION TIME

REDUCE SESSION TIME

REDUCE PHA FATIGUE

Benefits – Increased Risk Discovery

- Data-driven risk-based decisions
- Increase PHA Quality and Completeness
- Elimination of bias
- Break down silos between processing units across organization and industry

"Would you prefer 5 people versus 1,000 people and their perspectives & experience when looking at the Risk at your site?"

HELPING
you MAKE
your WORLD
a SAFER
PLACE

RISK ALIVE

THANK YOU

PLEASE FEEL FREE TO REACH OUT WITH ANY QUESTIONS
AT RMAI@RISKALIVE.COM

Asset Risk Ranking: A Faster and Effective Approach Towards High-Performance Asset Integrity Management

Shaiq Bashir
Senior Engineer, Instrument
Qatar Fertiliser Company
sbashir@qafco.com.qa

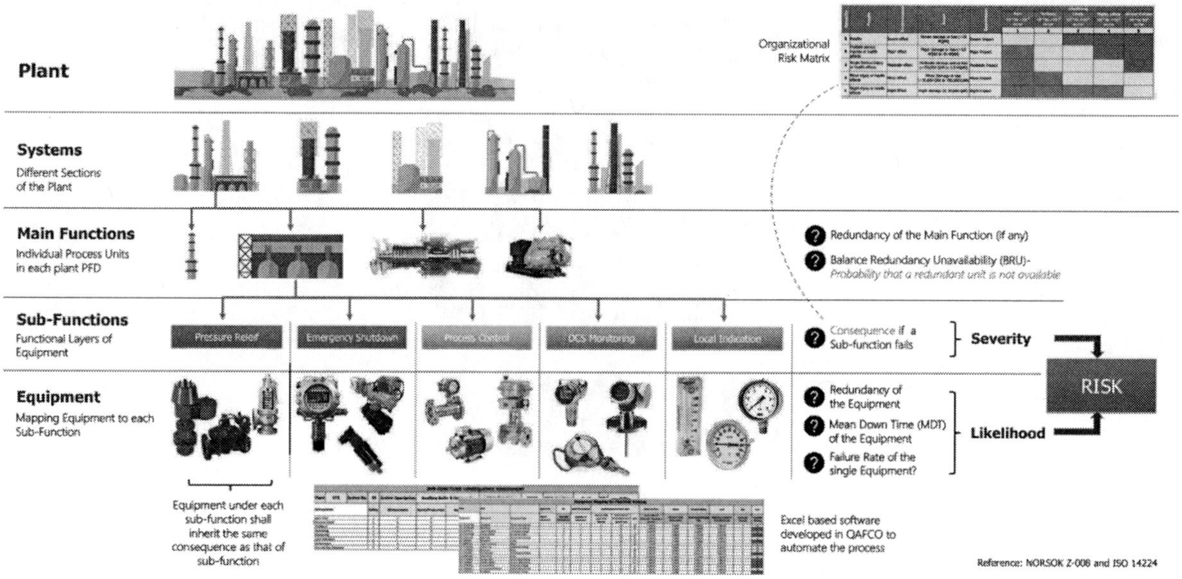

Cause analysis of hot work accidents based on text mining and deep learning

Hui Xu [a], Yi Liu [b], Lesong Zhao [a], Liang Jin [a], Yulin Ruan [a], Zhongxu He [a]
[a] College of Mechanical and Electrical Engineering, China University of Petroleum (East China), Qingdao 266580, China
[b] College of Chemical Engineering, China University of Petroleum (East China), Qingdao 266580, China

Introduction

Chemical industry
pillar industry

Hot work
unconventional work

Hot work accidents
severe consequences

- The large-scale accidents of chemical and hazardous chemicals in China, accidents involving hot work account for more than 50%.
- The accidents caused by hot work have shown an upward trend, and hot work has caused eight major and extraordinarily serious accidents since 2010.

Disadvantages of accidents' records analysis:

★ accidents are inadequate---not comprehensive
★ based on expert experience ---not comprehensive
★ manual processing excessively---time-consuming

Methodology

1.Topic recognition --- LDA
2.Cause classification automatically --- CNN
3.Key causes extraction ---Text mining
4.Cause visualization --- Word cloud

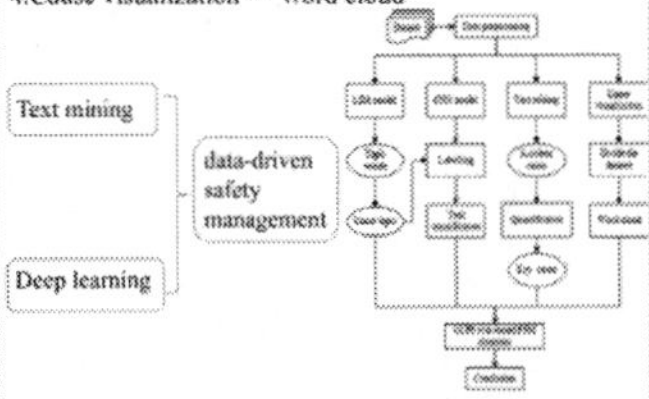

Fig. 1. Visualization of cause topics

Results

1.Topic recognition — LDA

Cause topics:

- No combustible gas detection or continuous monitoring
- Uncleaned combustibles
- Improper protection measures
- No special operation certificate or expired certificate
- Incomplete hazard identification
- Hot work approval is not in place

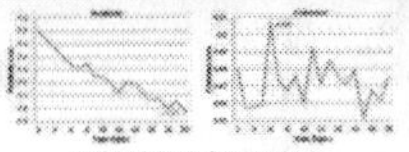

Fig. 2. Perplexity and coherence score curve

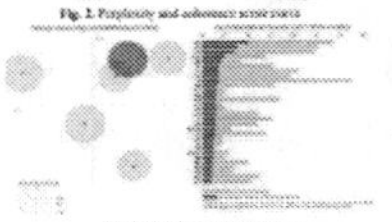

Fig. 3. Visualization of cause topics

2.Cause classification automatically --- CNN

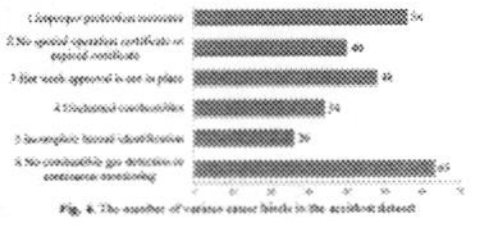

Fig. 4. The number of various cause labels in the accident dataset

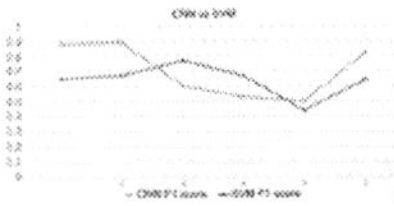

Fig. 5. Two classification model performance

3.Key causes extraction ---Text mining

$$C_i = \sum_{n=1}^{N} W_j$$

C_i is the importance score of the cause i,
N is the number of keywords that exist for this cause,
W_j is the *TF-IDF* value corresponding to the keyword j

No.	Keywords	No.	Key causes
C_1	no combustible gas detection or continuous monitoring	C_6	lack of safety education and training
C_2	chaotic safety management	C_7	improper protection measures.
C_3	work in violation of operating procedures	C_8	uncleaned combustibles
C_4	contractor qualification does not meet the requirements	C_9	incomplete hazard identification
C_5	no special operating personnel qualification certificate or certificate expired	C_{10}	improper emergency response

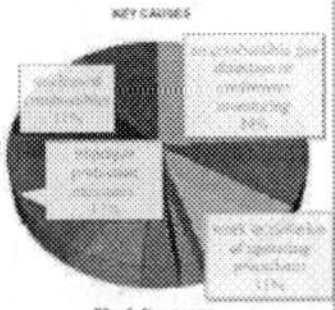

Fig. 6. Key causes

Fig. 5. Word cloud map of the top 20 words in different datasets. (General accident, b. Larger and above accidents)

Discussion

1.Most of the hot work accidents are caused by workers violating operating regulations. The specific manifestations are the first three key causes.

2.Controlling the first four key causes is more effective than a focus on hot work permits and the special operation certificate.

3.Emphasizing the importance and execution level of operating procedures by multi-frequency safety education and training.

Conclusion

1.Text CNN is a robust model than shallow ML models in predicting the cause.

2.Key causes of the hot work accidents:

★ no gas monitoring and continuous monitoring
★ uncleaned combustibles
★ improper protection measures
★ work in violation of operating procedures

3.The discovery of insufficient PSM

★ operating procedures
★ safe work practices,
★ operational readiness

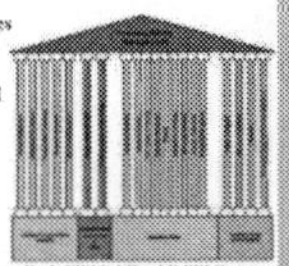

Fig. 8. CCPS Risk-Based 20 PSM Elements

Acknowledgments

The authors gratefully acknowledge the financial support from the 2019 Yankuang Science and Technology project (Grant no. YKZB 2020-173);
Key Research and Development Program of Shandong Province (Grant no. 2019JZZY020502).

Background

"Put the operator alarms at the boundaries of where you normally operate" – EEMUA 191

Process operation is only rarely considered in alarm management today. We'd like to move alarms limits from where they are now to this boundary.

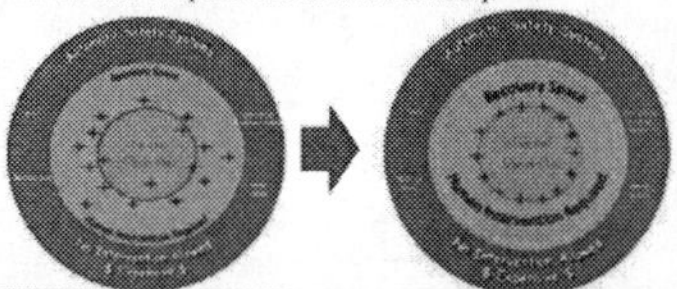

Improved Alarm Management with Big Data and Operating Envelopes

Alan Mahoney, PhD; Robin Brooks, PhD; John Wilson, PhD

Process Plant Computing Limited, Gerrards Cross, United Kingdom

PPCL

www.ppcl.com

Results

Avoidance of automatic safety systems and lost time is one measure of the true value of the alarm system. In this case, equipment downtime was reduced 20-fold from 2% to 0.1% following alarm rationalization.

2 years before ← → 2 years after

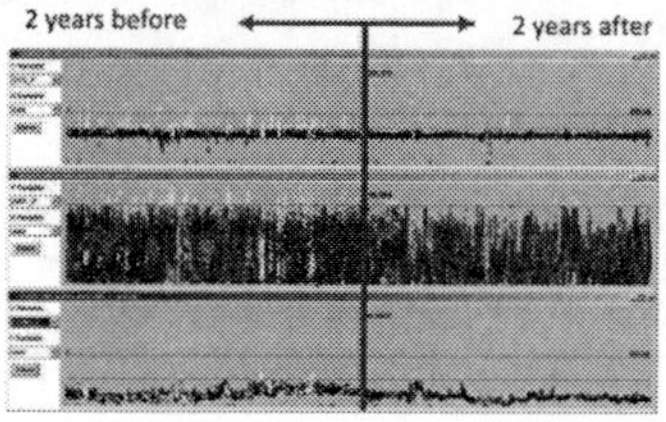

LNG Plant
- 4 trains, 8600 alarmed variables
- Team 1 Rationalization completed in 700 hours

Oil refinery
- 6 units, 3,600 alarmed variables
- 4 Unit Process Engineers
- Team 1 Rationalization in 320 hours

GTL plant
- 4,500 alarmed variables
- Team 1 Rationalization in 400 hours

Approach

Using the process operating envelope in the rationalization step and predicting alarm performance, better alarm limits can be obtained with far less work, also minimizing later change requirements.

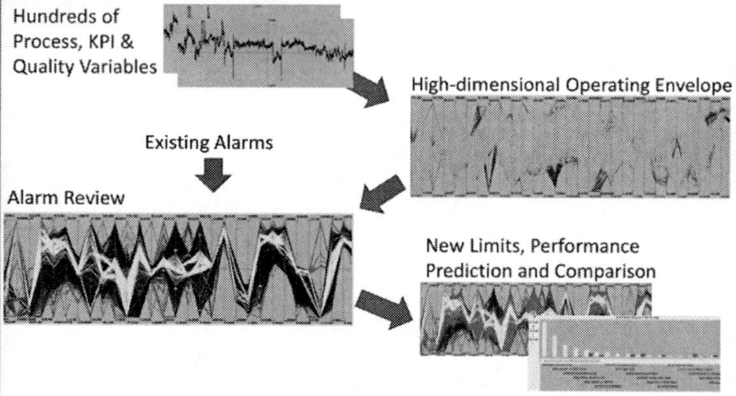

Benefits

- Reduced downtime
- Typical 80-90% predicted reduction in standing alarms and annunciation rate
- Significant manpower savings

Technology

Geometric Process Control
- Allows users to visualize datasets of thousands of variables
- Requires no advanced mathematics or statistics
- Captures fully multi-dimensional operating envelopes
- Supports hypothesis testing and what-if scenarios

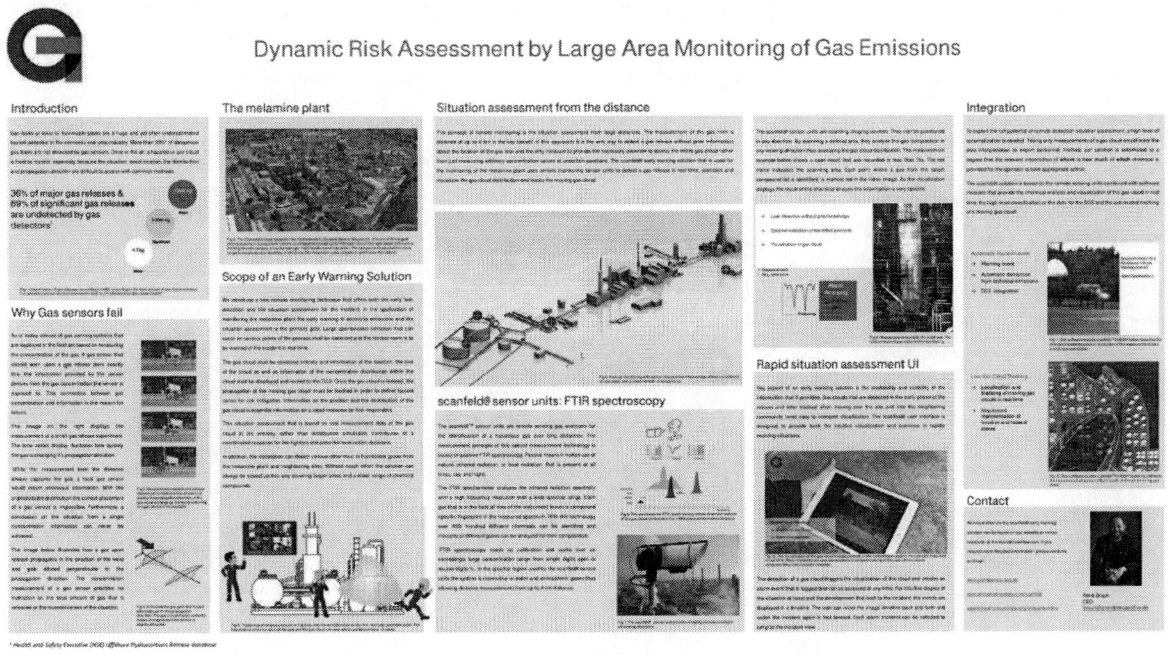
Dynamic Risk Assessment by Large Area Monitoring of Gas Emissions
Introduction
36% of major gas releases & 69% of significant gas releases are undetected by gas detectors
Why Gas sensors fail
The melamine plant
Scope of an Early Warning Solution
Situation assessment from the distance
scanfeld® sensor units: FTIR spectroscopy
Rapid situation assessment UI
Integration
Contact

Digitalization – What could possibly go wrong?

Roger Stokes FIChemE, BakerRisk Europe Ltd. rstokes@bakerrisk.com

Digitization, Digitalization , Big Data and Automation bring together new technologies, systems and data that allow humans and machines to interact with each other in ways that are continuously evolving. This brings with it the potential for major improvements including increased productivity, efficiency, innovation, safety and ultimately to improve the quality of life for everyone. This conference includes many excellent examples, but we must not forget that whenever new technology is introduced, so are new potential failure modes and types that are sometimes difficult to foresee. In high hazard industries, where the consequences of failure are severe, this could introduce unacceptable risks. This is particularly an issue where the potential negative interactions between these new ways of doing things and our existing systems, particularly human interactions and performance factors, are not fully understood and assessed. There are many historic examples where technical progress has led to incidents and much has been learnt to avoid similar failures; however as yet newer technologies and more complex systems are rapidly coming into play, there may be failure modes that are unknown or not properly understood. The use of existing process safety management tools can help but it is suggested that new technologies should be tried out on low risk applications first, in order to gain experience and identify bugs and failure modes before they are applied to high risk activities.

Some Change Management tools

Management of Change (MOC)	An analysis to identify, control / mitigate and document/ communicate **changes** in risks associated with the modification of equipment, raw materials, operating parameters, systems, procedures, organisations, etc.
Failure mode and Effect Analysis (FMEA)	A hazard identification technique in which all **known** failure modes of components or features of a system are considered in turn and undesired outcomes are noted. (CCPS)
Fault Tree Analysis (FTA)	A method used to analyze graphically the failure **logic of a given event**, to identify various failure scenarios and to support the probabilistic estimation of the frequency of the event.
Pre-start-up Safety Review (PSSR)	A systematic and thorough check of a process prior to the introduction of a highly hazardous chemical to a process.

But what is a change?

- Replacement in Kind (RIK)?

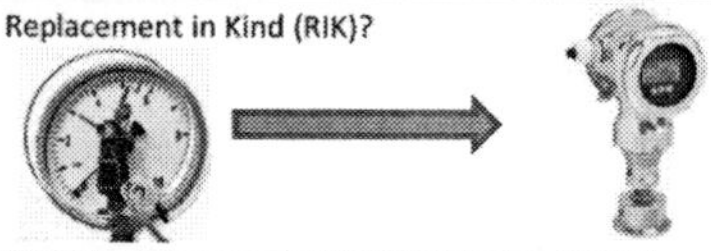

EXAMPLE - 6 months after first installation (1980s)

- Some high pressure alarms not working despite passing commissioning test
- Alarm set at full scale (or beyond) – OK with old analogue system
- Digital system 4-20mA – could not recognise data beyond instrument range (saturation)
- Simulated signal sent for commissioning and testing (not final element)

An unknown failure mode at the time

Known Failure Modes?

The unknown unknowns

We are not clever enough to predict all possible failure modes of the system - particularly when it comes to new technology, the human interface and complex systems

Why not try it out first?

Nuclear industry using wireless instrumentation 😲

Installed on accounting metering system to gain operating experience of reliability and failure modes. 👍

Case Studies

Incident:

- Not enough mud being pumped down an underbalanced well whilst drill pipe was removed ("Tripping")
- Pressure in well > head of mud
- Blowout and fire killed five workers

Cause:

- Many contributing factors
- New electronic version of a "trip sheet" - automatically calculated the fluid balance in the well, rather than relying on a manual calculation as done before
- Operator not trained in the use of the electronic trip sheet, **didn't understand it**
- Drillers had also turned off the alarm system that was giving excessive nuisance alarms, masking more critical alarms

Lessons:

- **New automated checklist system / inadequate training or understanding** of the new system was a key causal factor with this incident.
- **Management of Change** system needs to ensure **employees are involved** in the details of procedural changes and are properly trained prior to start-up.
- Would expect to be checked as part of a Pre-Start-up Safety Review (PSSR).

Boeing 737 MAX Crashes

Incidents:

- Lion Air, Indonesia 28 Oct 2018 – 189 fatalities
- Ethiopian Airlines, 10 March 2019 – 157 fatalities

Direct cause:

Newly installed control feature forced nose down and pilots unable to regain control.

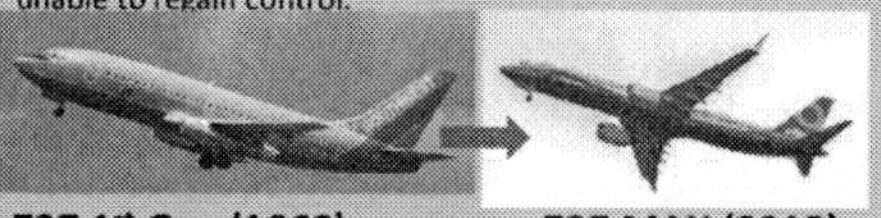

737 1st Gen (1968)	737 MAX (2018)
Basic mechanical / hydraulic controls	More sophisticated electronic systems; still controlling mechanical / hydraulic systems
Small diameter engines	Large diameter engines (more efficient)

- With new engines mounted higher and forward, the 737 MAX had a nose-up tendency under high power levels
- Automated system MCAS installed to overcome this tendency, so it feels to the pilot like previous generation.
  - Operates by pitching nose downwards.
- **No** pilot simulator training
- MCAS description not included in manuals
- Input from single angle of attack (AOA) sensor
- Erroneous reading from AOA sensor led to repeated activation of MCAS and nose-down inputs that pilots could not correct.

Lessons:

- Design change led to **new failure mode** – apparent failure to consider this
- Reliance on single AOA sensors (2 were fitted)
- Repeated activation that **pilot could not cancel** without isolating flight trim system
- Inadequate **training and understanding** of new system
- **Management and culture failings**

Pipework Integrity Monitoring

Incident:

- Major Loss of containment when pipe ruptured leading to multiple fatalities

Cause:

- Gradual reduction in pipe wall thickness until pressure could not longer be contained
- Spot measurement/ continuous monitoring at key Thickness Measurement Locations (TMLs) were carried out
- Data used to predict retirement dates using software/ algorithms and comprehensive reporting
- 13th year of operation, the thickness checks/ software calculation showed 9 years to go before the retirement thickness reached (total life 22 years)
- Pipe failed after 17 years of service
- **Incorrect data had been selected in the software**
- **Automated calculation of retirement date used corrosion rates from early in the life of the pipe, when rates were much lower.**
- **Use of most recent corrosion rates would have identified the problem**
- **Maintenance planners did not understand the software**

Lessons:

- All staff to have a full **understanding of new automated systems**
- These can also introduce potential **new modes of failure.**
- **Human factors associated with the new system** not adequately considered

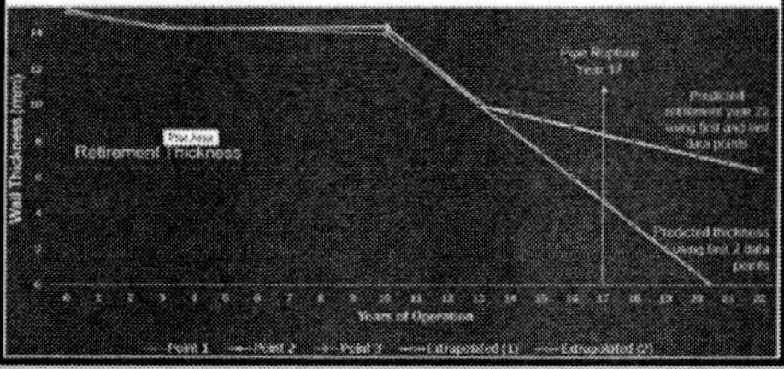

Human vs automated systems

- Are we replacing a human safety barrier with something more or less reliable?
- What will the human do if/ when the system fails to work?
  - Is it understood / too complex?
  - Will the human know what to do when the automated system cuts out?
- Are we transferring an element of a safety critical activity to someone else and are there adequate procedures in place to control this?
  E.g. - Labelling process equipment using bar codes / NFC
  - End of life prediction via TMLs
- Human Factors in Risk Assessment

Where technological change appears to provide additional safety barriers/ layers of safety, we must ensure that we are not degrading our existing barriers due to factors associated with human performance.

Conclusions

- Tempting to rush to adopt new tools and technologies
  - *Significant benefits*
- Overall analysis required to assess impact on safety critical systems
  - *Many of which will involve human factors – people/ equipment / technology*
- Use a systems approach for process safety
  - *The big picture, looking at whole systems, not just units in isolation*

Make sure the use of the newer technology is of benefit and does not have an adverse effect on the overall safety of the system

The Development of an Automatic Risk Scenario Classifier using an Artificial Intelligence Approach in a Petrochemical Plant

Layane Mirela Ornelas Sacramento Oliveira – layane.mirela@ufba.br| Delano Mendes de Santana – delano.mendes@ufba.br | Narciso Ruzzarin – narciso.ruzzarin@braskem.com

INTRODUCTION

In order to facilitate HAZOP studies in a Petrochemical plant, this work uses the Knowledge Discovery from Data (KDD) to develop an automated risk scenario classifier, according to pre-established risk classification criteria.

DEVELOPMENT

Database

- Organized historical risk scenario from a polyethylene plant into a database with some attributes related to HAZOP

Data Mining

- Used Weka Workbench to apply Machine Learning algorithms to evaluate the influence of attributes in risk classification and predict it
- Developed an easy-to-understand explicit classification tree validated by Process safety specialists (Accuracy = 87%; Kappa = 0.79)

Interface

- Used Microsoft Power Apps to develop an user interface to support risk analysis and maintain continuous learning of algorithm.
- Users choose the attributes through checkboxes and the risk classification is given. If the classification is done wrongly by the model, the user can inform the correct classification.

MAIN RESULTS

- Support less experienced engineers, reducing the differences in interpretations, effort, and time of HAZOP studies
- Support operational team in risk analysis to release process safety barriers and simpler change management;
- Excellent tool for initial risk analysis that can be adapted and applied in other processes with different criteria for risk classification.

FUTURE WORK

- Further studies to extract more knowledge about risk scenarios
- Mapping possible causes and consequences for a process and suggest recommendations
- Predict risk scenarios in other plants

USER INTERFACE

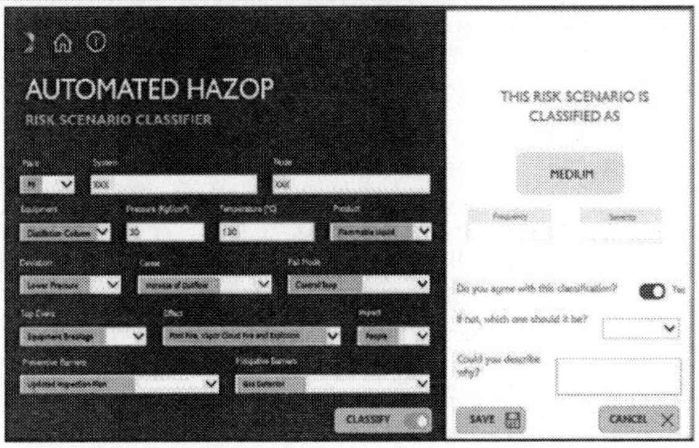

REFERENCES

Single, J. I., Schmidt, J. & Denecke, J. State of research on the automation of HAZOP studies. J. Loss Prev. Process Ind. 62, 103952 (2019)

Fayyad, U., Piatetsky-Shapiro, G. & Smyth, P. From Data Mining to Knowledge Discovery in Databases. AI Magazine vol. 9078 637–648 (1996)

Witten, I. H., Frank, E. & Hall, M. A. Data Mining - Pratical Machine Learning Tools and Techniques. Morgan Kaufmann (2008).

Palmer, T. Microsoft PowerApps as an Alternative Solution for Business Application Development. (Haaga-Helia University of Applied Sciences, 2020)

The University Of Sheffield.

Aaron S. Yeardley, Jude Ejeh, Joan Cordiner, Solomon F. Brown*

Department of Chemical & Biological Engineering, University of Sheffield, Sheffield, S1 3JD

*Corresponding author. E-mail address: s.f.brown@sheffield.ac.uk

An investigation into an ensemble of machine learning techniques to provide an optimum maintenance schedule

Introduction

Machine maintenance is paramount to the process industry. The operating state of an industrial process is very dependent on a business having effective maintenance procedures. Therefore, effective planning of maintenance is essential.

- Predictive maintenance has the potential to improve manufacturing capabilities.
- Maintenance time estimation could prevent rushed maintenance an serious process safety incidents.
- Maintenance optimisation can produce a schedule to increase profits.

Objectives

- Build such a methodology that will implement a maintenance policy that produces an optimum maintenance schedule.
- Investigate each step in the methodology using data provided by a Fischertechnik (FT) factory model (Klein et al., 2019):
  - Which classification methods can successfully predict faults?
  - Which regression methods can increase time estimation accuracy?
  - How beneficial can an optimisation schedule be?

Proposed Maintenance Policy Methodology

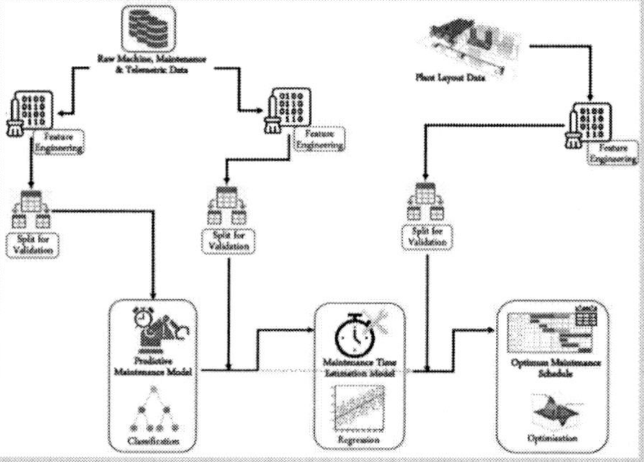

Results

Predictive Maintenance

This research analysed five machine learning classification techniques and evaluated each using three popular classification error diagnostics.

The Quadratic Discriminant Analysis achieved the best results.

Therefore, we are confident the machine fault can be successfully predicted for the requirements of this policy.

Table 4: Resulting diagnostic values from the predictive maintenance

	Precision	Recall	F1 Score
Decision Tree	0.839	0.878	0.838
Random Forest	0.735	0.858	0.792
Neural Network	0.788	0.861	0.799
AdaBoost	0.736	0.858	0.793
Quadratic Discriminant Analysis	0.880	0.883	0.877

Results

Time Estimation Model

The Gaussian Process (Yeardley et al., 2021) achieved the best results and the residuals were plotted below.

Results show ML provide a significant benefit to maintenance models.

Table 5: Resulting diagnostic values from the time prediction

	R squared	Standardised RMSE	RMSE (hours)
Gaussian Process	66.3%	0.600	0.202
Neural Network	57.9%	0.670	0.226
Gradient Boosting	56.9%	0.681	0.230
Support Vector Regression	51.8%	0.723	0.244
Random Forest	51.7%	0.720	0.243

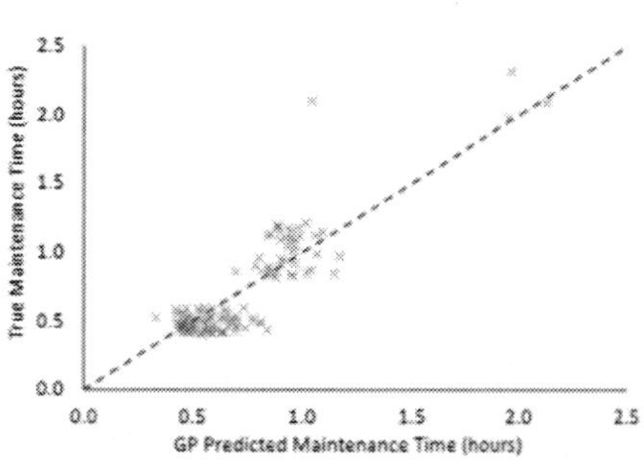

Figure 3: The residuals predicted by the Gaussian Process regression model.

Results

Optimisation Schedule

Mathematical optimisation minimised the costs given system constraints. The data was split into three cases based on calendar dates. The optimisation model was implemented using our proposed policy and compared to condition based scheduling.

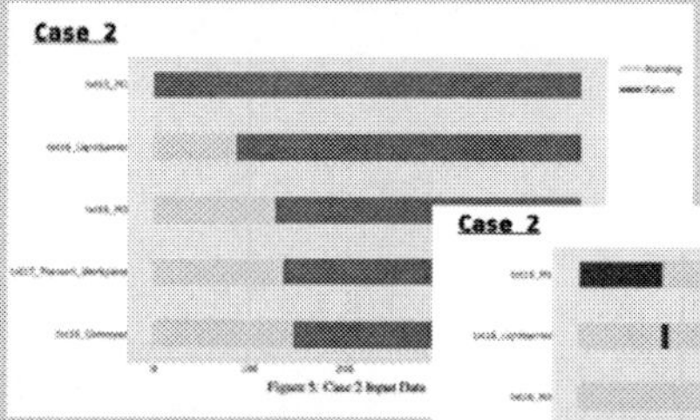

Figure 5: Case 2 Input Data

Case 2 was the most complex day with five machines all incurring faults early on.

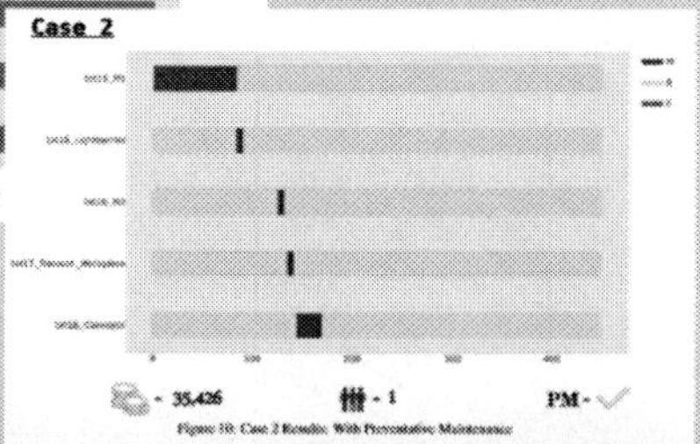

Figure 10: Case 2 Results: With Preventative Maintenance

Our proposed strategy saved:

£119,664

By saving on downtime costs and reduced maintenance times.

Conclusion

The technical contributions can be summarised as:

- We have investigated various machine learning tools for process safety.
- Tested a novel maintenance policy as a proof of concept showing machine learning can be used to improve maintenance scheduling.

Klein, P., Bergmann, R., 2019. Generation of complex data for AI-based predictive maintenance research with a physical factory model. ICINCO 2019 - Proceedings of the 16th International Conference on Informatics in Control, Automation and Robotics 1, 40-50. doi:10.5220/0007830700400050.

Yeardley, A.S., Bellinghausen, S., Milton, R.A., Litster, J.D., Brown, S.F., 2021. Efficient global sensitivity-based model calibration of a highshear wet granulation process. Chemical Engineering Science 238, 116569. URL: https://doi.org/10.1016/j.ces.2021.116569, doi:10.1016/j.ces.2021.116569.

Getting to Zero Errors through Procedure Clarity and effective Knowledge Transfer.

VOOVIO

The Industry Challenge:

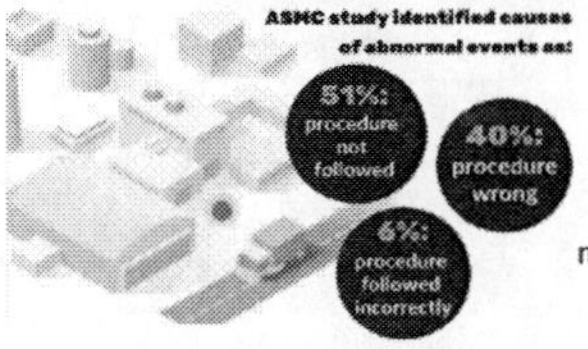

'Failure to follow procedure' is often cited as root cause in incident investigations. However, Voovio data shows 'Procedure Clarity' is a major source of error overlooked in the process industry.

3 reasons for suboptimal SOP clarity:

1. **Procedure does not reflect operator practice:** Written procedures vary from the unit's established practice
2. **Ambiguities, generalizations & technical content in procedures:** assume clear understanding by the operator
3. **Inexperience; Lack of practice:** Operators only experience or practice ~40% of procedures they qualify on for any given job

What is Procedure Clarity for an Operator?

1. Clear and concise actions
2. Ability to visualize procedure steps in field
3. Ability to practice standardized execution *(learn by doing)*

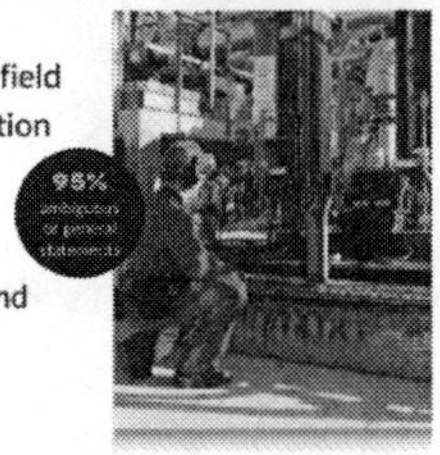

Voovio analyzed over 300 customer Standard Operating Procedures and found that 95% of them contain ambiguous or general statements.

Customer example:

1 step in an SOP, which requires the operator to perform 53 individual actions in the field:

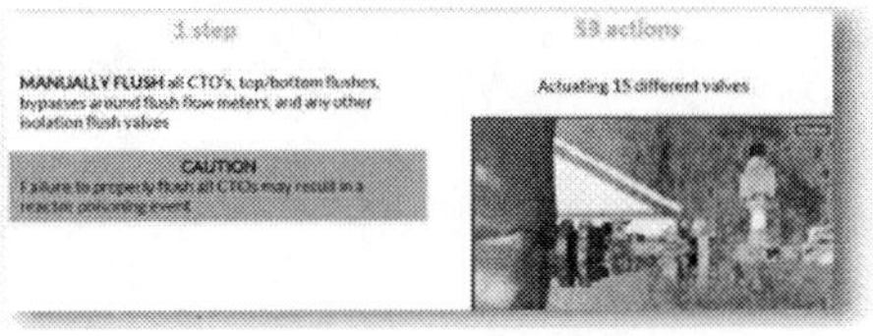

For more information, contact info@voovio.com. Or visit us on voovio.com & LinkedIn @Voovio.

Knowledge Automation. Made Easy.
voovio.com

VOOVIO

The Solution:

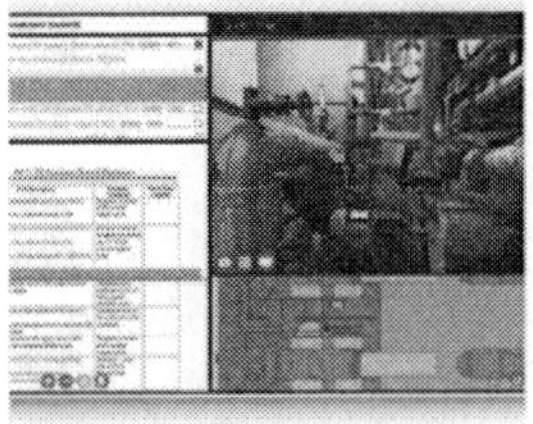

By utilizing procedure simulators in a Knowledge Automation platform you remove ambiguity and provide clear, concise action steps that operators can learn and practice 24/7 without the need of a trainer or SME. This system captures tribal knowledge and makes it easily accessible.

What do operators think of Procedure Simulators?

In surveys conducted with over 100 end users between 2018-2021, 100% of respondents "Strongly Agreed" Procedure Simulators:

- make it easier to understand procedures than their current (traditional) training methods
- provide clarity on aspects of their job that they did not have before
- increase compliance with operating procedures in their area because:
  - Visual clarity
  - Procedures are accurate and action-based
  - Procedures and practices are standardized

Knowledge Automation Cycle:

In addition, a Knowledge Automation platform's testing capabilities provide metrics to track your organization's know-how of procedures, enabling you to identify gaps and predict human reliability issues ahead of time.

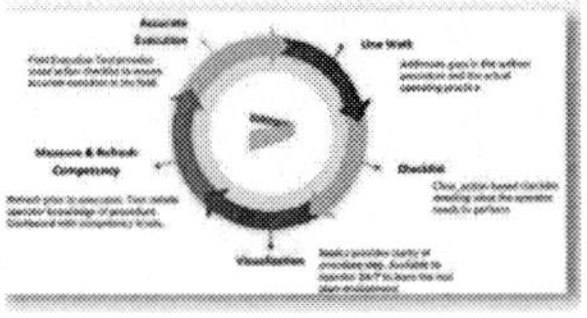

For more information, contact info@voovio.com. Or visit us on voovio.com & LinkedIn @Voovio.

Knowledge Automation. Made Easy.
voovio.com

Operational Sustainability, LLC

2021 European Conference on Process Safety & Big Data

Driving Operational Excellence™

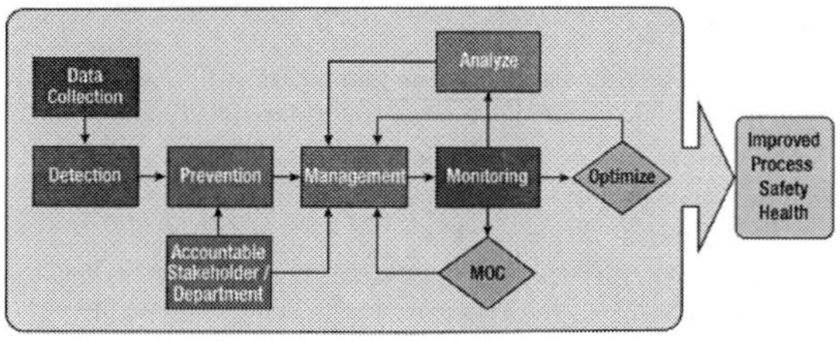

CCPS	Lagging	Leading
Commit to Process Safety – How do we measure organization and community involvement?	Near miss rate or high potential incident rates – number of near misses or high potential incidents per month	Number of employees involved in management or supervisor engagement sessions per month to discuss process safety and key drivers to performance
Understanding Hazards and Risks – What is an organization's understanding of the hazards and risks?	Spill rate – % of LOPCs as a result of a human performance issue	% of incidents addressed as identified in PHA, operating procedures, or Integrity Operating Windows
Manage Risk – Are systems effective to reduce risks and minimize incidents?	Leak rate – % of LOPCs as a result of a mechanical integrity failure	Potential MI failure – % of piping circuits inspected in last 12 months and % identified within 20% of minimum thickness
Learn from Experience – Is constant improvement occurring?	Tier 1 or 2 process safety event rate – number of Tier 1 or 2 events per [illegible] base	% of incident investigation root causes communicated to operations that resulted in a change in procedure or action incorporated into routine

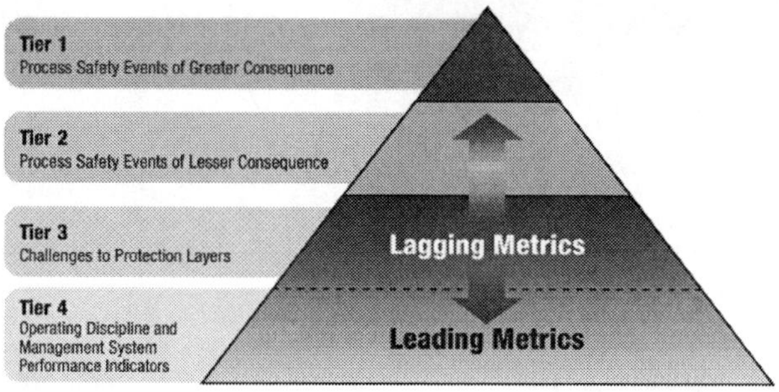

David Drerup
CEO – Operational Sustainability, LLC
Info@DrivingOE.com
713.355.2900
www.DrivingOE.com

INGENERO

Excellence Through Insight

CCPS : Big Data Analytics on Process Safety

Confidential

Deviation Tracker

INGENERO
Excellence Through Insight

Key Features of Deviation Tracker

- Alerts deviation which may lead process safety incident
- Based on 14 Days of Operation
- More than 4000 Instrumentation covered across 28 equipment
- Applied to major equipment like Furnaces, Compressors, Columns

Tracks multiple variables concurrently
Monitoring plant stability

Stability Index
Aggregated representation of instrumentation health

Automatic alerts
Triggers alerts for abnormal deviations

Smart recognition
Distinguishes between operational changes and deviations

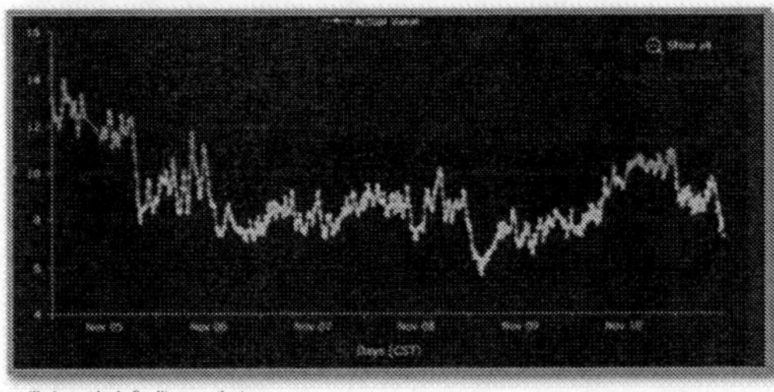

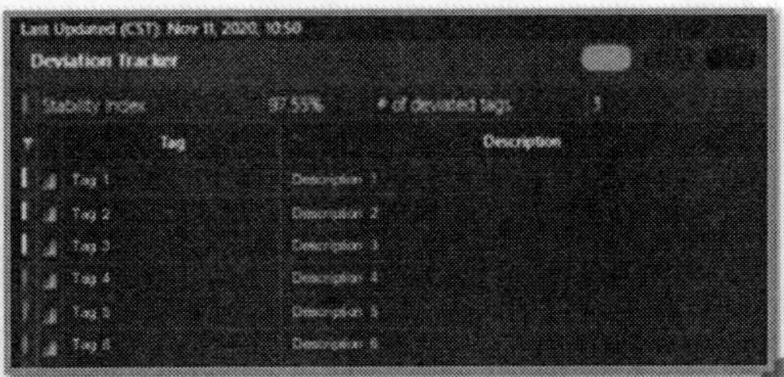

Confidential & Proprietary

2

AUTHOR INDEX

AIChE
120 Wall Street, FL 23
New York, NY 10005-4020

ISBN 978-1-7138-6321-2

CURRAN ASSOCIATES INC.
proceedings.com